I0033031

United States SPEE, World's Engineering Congress

Engineering Education

being proceedings of Section E of the World's Engineering Congress - Vol. 2

United States SPEE, World's Engineering Congress

Engineering Education
being proceedings of Section E of the World's Engineering Congress - Vol. 2

ISBN/EAN: 9783337255992

Printed in Europe, USA, Canada, Australia, Japan

Cover: Foto ©berggeist007 / pixelio.de

More available books at **www.hansebooks.com**

Engineering Education.

Proceedings of the Second Annual Meeting

OF THE

Society

FOR THE

Promotion of Engineering Education

HELD IN

Brooklyn, New York, August 20-22, 1894,

IN CONJUNCTION WITH THE

American Association for the Advancement of Science.

Volume II.

EDITED BY

Geo. F. Swain, Ira O. Baker, J. B. Johnson.

COMMITTEE.

Columbia, Mo.:
E. W. STEPHENS, PRINTER
1895.

OFFICERS

of the

Society for the Promotion of Engineering Education.

1894-5.

President—GEORGE F. SWAIN, Mass. Inst. Tech., Boston, Mass.

Vice-Presidents—ROBERT H. THURSTON, Ithaca, N. Y.
F. O. MARVIN, Lawrence, Kan.

Secretary—J. B. JOHNSON, Wash'n Univ'ty, St. Louis, Mo.

Treasurer—STORM BULL, U. of W., Madison, Wis.

MEMBERS OF THE COUNCIL.

Terms of Office Expire in 1895.

H. T. BOVEY, Montreal, Can.
W. H. BURR, New York City.
O. H. LANDRETH, Schenectady, N. Y.
MANSFIELD MERRIMAN, Bethlehem, Pa.
WM. G. RAYMOND, Troy, N. Y.
GEO. F. SWAIN, Boston, Mass.
DEVOLSON WOOD, Hoboken, N. J.

Terms of Office Expire in 1896.

I. O. BAKER, Champaign, Ill.
STORM BULL, Madison, Wis.
S. B. CHRISTY, Berkeley, Cal.
J. GALBRAITH, Toronto, Can.
J. B. JOHNSON, St. Louis, Mo.
F. O. MARVIN, Lawrence, Kan.
C. D. MARX, Palo Alto, Cal.

Terms of Office Expire in 1897.

L. S. RANDOLPH, Blacksburg, Va.
H. T. EDDY, Minneapolis, Minn.
J. J. FLATHER, Lafayette, Ind.
ALBERT KINGSBURY, Dunham, N. H.
S. W. ROBINSON, Columbus, Ohio.
J. P. JACKSON, State College, Pa.
ROBERT H. THURSTON, Ithaca, N.Y.

TABLE OF CONTENTS.

BRIEF SUBJECT INDEX.

CONSTITUTION

OF THE

Society for the Promotion of Engineering Education.

1. NAME.—This organization shall be called the SOCIETY FOR THE PROMOTION OF ENGINEERING EDUCATION.

2. MEMBERS.—Members of this Society shall be those who occupy, or have occupied, responsible positions in the work of Engineering instruction, together with such other persons as may be recommended by the council.

The name of each candidate for membership shall be proposed in writing to the council by two members to whom he is personally known. Such names, if approved by the council, shall be voted on by the Society at the annual meetings.

3. COUNCIL.—The council of this Society shall consist of twenty-one members chosen from the Engineering Schools of the United States and Canada, and one-third of the council shall retire annually.

Any member of this Society shall be eligible to election to the council, provided that not more than one member of the council shall be from any one school.

Members of the council shall be elected by ballot by the Society at its annual meeting.

The council shall constitute a general executive body of the Society, pass on proposals for membership, attend to all business of the Society, receive and report on propositions for amendments to the constitution, and shall have power to fill temporary vacancies in the offices.

4. OFFICERS.—There shall be a president, two vice-presidents, a secretary, and a treasurer, each to hold office for one year. They shall be chosen by vote of the members at the annual meeting. Members of the council only shall be eligible to these offices.

5. FEES AND DUES.—The admission fee shall be three dollars and the annual dues two dollars, payable at the time of the annual meeting.

Those in arrears more than one year shall not be entitled to vote nor to receive copies of the proceedings, and such members shall be notified thereof by the secretary one month previous to the annual meeting.

6. MEETINGS.—There shall be a regular meeting occurring at the time and place of the meetings of the American Association for the Advancement of Science, or of some one of the National Engineering Societies, or otherwise as the council may determine.

7. PUBLICATIONS.—The proceedings of the Society, and ·such papers or abstracts as may be approved by the council, shall be published as soon as possible after each annual meeting.

8. AMENDMENTS.—This constitution may be amended by a two-thirds vote at any regular meeting, after action thereon by the council.

Rules Governing the Council.

First. The officers of the Society shall constitute a committee to arrange the time and place of the annual meeting, and also to prepare a program for the same.

Second. The reading of papers shall be limited to fifteen minutes each, and abstracts of the same of about 300 words or less shall be printed when practicable and distributed in advance to the members.

Third. The time occupied by each person in the discussion of any paper shall not exceed five minutes.

Fourth. There shall be but one session daily for the reading of papers and for discussions.

OBJECTS AND OPPORTUNITIES

Society for the Promotion of Engineering Education.

This Society was organized at the close of the Engineering Congress held in Chicago in August, 1893. It grew out of a common feeling of the seventy or more members of Section "E" of that Congress on Engineering Education that a perminent organization should be established of those engineers and teachers most interested, which should meet annually for conference and discussion.

The first annual meeting was held in Brooklyn in August, 1894, in connection with the American Association for the Advancement of Science. Over 100 members were present at that meeting, and the total membership was raised to 156. The proceedings of Section "E" of the World's Congress was published in book form by this Society, making an octavo volume of 342 pages, which constitutes Volume I. of the proceedings of this Society.

Future meetings will be held in conjunction with various engineering society meetings or with other educational bodies.

While this organization already includes in its membership a large number of the leading educators in the Engineering schools of this country, it is desirable, in order that the Society may completely fulfill its mission, that practically all the professors of Engineering in this country, as well as those practicing engineers who are interested in the development of our engineering schools should become members of this Society, and contribute to its proceedings.

The Engineering schools of America are still in a formative stage, with a prospect that many of them will become, at a very early date, very nearly the ideal Engineering schools.

The shaping of the technical education of a nation is a work which may well call for the continuous service of our ablest educators, and the advantages of participating in the work should accrue to each and every school, college, or university in this country, which undertakes to impart technical instruction, through its professors and teachers who are members of the Society which is principally charged with this work. There is now little question that this Society will have a great and lasting influence in shaping the development of our Engineering schools, and while the teachers of our leading technical schools have, to date, shown the greatest interest in this work, the greatest advantages are likely to be reaped by those members who represent the newer and smaller schools, the principal development of which is still in the future.

Persons desiring to become members of this Society can obtain blank forms of application and other information, on application to the secretary of the Society.

PROCEEDINGS.

MORNING SESSION, 10:45 A. M.

The second annual meeting of the SOCIETY FOR THE PROMOTION OF ENGINEERING EDUCATION was held in the Polytechnic Institute Building of Brooklyn, New York, in conjunction with the meetings of the American Association for the Advancement of Science. The meeting was called to order at 10:45 A. M. by President DeVolson Wood.

The secretary announced the order of business, and presented a list of names recommended by the council for membership, all of which were duly elected, and are included in the list of members given on subsequent pages.

The secretary presented his report which was received and filed.

The treasurer presented a report which was referred to the council for auditing.

On motion of Professor Merriman a committee of five was appointed by the president to make nominations for filling the positions on the council made vacant by the expiration of terms of office. Wm. G.

(1)

Library
N. C. State College

Raymond, Ira O. Baker, C. Frank Allen, H. S. Jacoby, and J. H. Kinealy were appointed such committee.

The secretary announced that, at the request of the officers of the Society, John Wiley & Sons had sent sample copies of all their engineering publications, and had placed them on exhibition in a room in the building where an assistant would be in constant attendance.

The president then read his annual address, after which the program for the afternoon was announced and the session adjourned.

AFTERNOON SESSION, 2:50 P. M.

The Society was called to order with Vice-President George F. Swain in the chair.

Professor F. O. Marvin, of the University of Kansas, presented a paper on "Common Requirements for Admission to Engineering Courses." This paper had been printed in advance and placed in the hands of the members. After a lengthy discussion on the above paper, Professor Wm. S. Aldrich, of the University of West Virginia, presented a paper on "Engineering Education and the State University." After the reading of this paper, Dr. Charles E. Emery was introduced as representing the local committee of entertainment, and offered his services in directing or accompanying the members of the Society to the local points of interest.

Professor Palmer C. Ricketts, Director of the Rensselaer Polytechnic Institute of Troy, N. Y., then opened the discussion on the subject of "Graduate and Post Graduate Degrees" by reading a paper. This was followed by a paper on the same subject by Professor George F. Swain, of the Mass. Inst. Tech.

At this point an invitation extended by Mr. Emery to inspect the operation of welding together street railway rails by electricity was accepted, and the Society adjourned.

TUESDAY, AUGUST 21, 1894.

MORNING SESSION, 10:45 A. M.

The meeting was called to order by Vice-President George F. Swain, and a second list of names recommended for membership by the council were balloted for.

The committee on nominations of new members of the council submitted the following: H. T. Eddy, J. J. Flather, J. P. Jackson, A. Kingsbury, L. S. Randolph, S. W. Robinson and R. H. Thurston, which were duly elected.

The committee on nominations for the council was continued as a committee on nominations of officers for the coming year.

Dr. Robert H. Thurston, of Cornell University, then continued the discussion on "Graduate and Post Graduate Degrees," after which the subject was discussed by various members.

Professor Mansfield Merriman read a paper on "Teachers and Text Books in Mathematics for Engineering Students," which was generally discussed, and the session adjourned.

Vice-President Swain in the chair.

Professor J. B. Johnson read a paper on "Teaching Engineering Specifications and the Law of Contracts," which was followed by a discussion upon the same.

A paper on "Mechanical Drawing and Lettering in Engineering Schools," by Professor J. J. Flather, was read and discussed.

Professor Storm Bull, of Madison, Wis., then read a paper entitled, "Some German Schools of Engineering," after the discussion of which Professor Lanza, of the Mass. Inst. Tech., presented a paper on "The Organization and Conduct of Engineering Laboratories, and the Equipment of the Laboratories at the Mass. Inst. Tech."

A paper by Professor C. M. Woodward on "Early Instruction in Physics and Mechanics," and one by Professor C. H. Benjamin on "Text Books Considered as Such and not as Works of Reference," were read by the secretary, after which the meeting adjourned.

WEDNESDAY, AUGUST 22, 1894.

MORNING SESSION, 10:30 A. M.

Vice-President Swain in the chair.

After the election of the new members reported by the council, the committee appointed to examine the treasurer's books reported that they had performed that duty, and found the accounts correct.

The council recommended to the Society the following rules for the government of future meetings:

First. The officers of the Society shall constitute a committee to arrange the time and place of the annual meeting, and also to prepare a program for the same.

Second. The reading of papers to be limited to fifteen minutes each, and that abstracts of the same of about 300 words or less be printed when practicable and distributed in advance to the members.

Third. The time occupied by each person in the discussion of any paper shall not exceed five minutes.

Fourth. There shall be but one session daily for the reading of papers and for discussions.

These rules were adopted by the Society.

The nominating committee reported the following names for officers for the ensuing year: President, George F. Swain; vice-presidents, Robert H. Thurston, F. O. Marvin; secretary, J. B. Johnson; treasurer, Storm Bull, all of whom were duly elected.

In reporting this ticket the committee wished it to

be understood that they did not wish to establish a precedent in favor of a continuous term of office of secretary and treasurer, but that in the infancy of the organization they thought it wise to continue these officers in office for another year.

It was moved by Professor Landreth that a committee of five be appointed by the chair to consider and report to this Society at its next meeting upon the subject of "Entrance Requirements for Engineering Schools," and that the report be printed and distributed to the members before the next meeting. After discussion this motion was adopted, and the president appointed the following as such committee: F. O. Marvin, Mansfield Merriman, J. J. Flather, H. W. Tyler and J. P. Jackson.

AFTERNOON SESSION, 2:30 P. M.

Vice-President Swain in the chair.

The Society met for the reading of papers in two sections, one taking up those papers more nearly related to civil engineering and the other those related to mechanical engineering. The time being so limited, the papers given to the latter section were read by title only. In the civil engineering section a paper was read by Professor Porter, of the Mass. Inst. Tech., on "Laboratory Work and Equipment in a Civil Engineering Course."

Professor Stout's paper upon "Some Mistakes in the Management of College Field Practice" was read by the secretary.

Professor Marburg, of the University of Pennsylvania, read a paper upon "The Teaching of Structural Engineering," and Professor Allen, of the Mass. Inst. Tech., one on the "Education of Civil Engineers for Railroad Service." Some discussion was had upon the above papers, after which the two sections united for a final business session.

The constitution was amended on recommendation of the council to read as given in this volume.

On motion, the thanks of the Society were tendered Dr. Emery, the authorities of the Brooklyn Polytechnic Institute, and the local committee for courtesies extended to the Society.

On motion, the thanks of the Society were tendered to President DeVolson Wood for his excellent annual address, and for the interest shown by him in meeting with the Society, though suffering from illness.

On motion, the thanks of the Society were tendered to Messrs. John Wiley & Sons for the convenient display of their publications, which had been a great accommodation to the members of the society.

Adjourned.

J. B. JOHNSON, Secretary.

LIST OF MEMBERS.

NAME.	TITLE.	DATE OF MEMBERSHIP.
Adams, C. A., Jr	Instr. Elec. Eng. Harvard University, Cambridge, Mass...................................	1894
Aldrich, William S...........	Prof. Mech. Eng., University of West Virginia, Morgantown, W. Va.........................	1893
Allen, C. Frank..............	Assoc. Prof. Railroad Eng., Mass. Inst. Tech., Boston, Mass...........................	1893
Anderson, F. P......	Prof. Mech. Eng., State College of Kentucky, Lexington, Ky.......................... ..	1894
Ayer, A. W.............	Prof. Mech. Eng., University of Vermont, Burlington, Vt.	1894
Baker, Ira O.................	Prof. Civ. Eng., University of Illinois, Champaign, Ill.............	1893
Baldwin, Ward........	Prof. Civ. Eng., University of Cincinnati, Cincinnati, Ohio......	1893
Barbour, V. G................	Prof. Mech. & Bridge Eng., Dean Eng. Dept., University of Vermont, Burlington, Vt.......	1894
Barney, S. E., Jr............	Instr. Civ. Eng., Sheffield Sci. Sch. of Yale University, New Haven, Conn...............	1894
Beardsley, Arthur...........	Prof. Engineering, Swarthmore College, Swarthmore, Pa.............................	1893
Benjamin, Charles H.........	Prof. Mech. Eng., Case School of Applied Science, Cleveland, O.	1893
Bird, W. W......	Asst. Prof. Mech. Eng., Worcester Polyt. Inst..	1894
Bissell, G. W................	Prof. Mech. Eng., Iowa Agricultural College, Ames, Iowa.............	1894
Bovey, H. T..................	Prof. Civ. Eng. and Applied Mechanics, M'Gill University, Montreal, Canada...............	1893
Bray, C. D...................	Prof. Civ. & Min. Eng., Tuffts College, Mass..	1894
Breckenridge, L. P...........	Prof. Mech. Eng., University of Illinois, Champaign, Ill.	1893
Brill, G. M................ ...	Experimental Engineer, in charge Syracuse evening drawing school, Syracuse, N. Y......	1894
Brown, Charles C......	City Engineer, Indianapolis, Ind...............	1894
Brush, Charles B.............	Ch. Eng. and Supt. Hackensack Water Co., Hoboken, N. J.	1893
Bull, Storm................ ...	Prof. Steam Eng., University of Wisconsin, Madison, Wis	1893

LIST OF MEMBERS.—Continued.

NAME.	TITLE.	DATE OF MEMBERSHIP.
Bunte, Dr. Hans..............	Aulic Counsellor, Professor in the Polytechnic School, Director of the Institute of Chemistry and Technology, and of the grand-ducal Bedensian experimental station at Karlsruhe.	1893
Burr, William II..............	Prof. Civ. Eng., Columbia College, New York City, N. Y.................	1893
Burton, A. E......	Prof. of Geodesy, Massachusetts Institute of Technology, Boston, Mass...............	1893
Carpenter, R. C..............	Assoc. Prof. of Exp. Eng., Sibley College, Cornell University, Ithaca, N. Y..............	1893
Carson, W. W................	Prof. Civ. Eng., University of Tennessee, Knoxville, Tenn...............................	1894
Christy, Samuel B............	Prof. Mining and Metallurgy, University of California, Berkeley, Cal...............	1893
Colburn, L. C.............. ...	Prof. Mech. Eng. and Mathematics, University of Wyoming, Laramie, Wy	1894
Cooley, M. E.................	Prof. Mech. Eng., University of Michigan, Ann Arbor, Mich.........................	1894
Coxe, Eckley B...............	Pres. Cross Creek Coal Co., Coxe Iron Mfg. Co., Drifton, Pa....	1894
Crandall, C. L................	Assoc. Prof. Civ. Eng., Cornell University, Ithaca, N. Y........	1893
Creighton, W. H. P. (U.S.N.)	Prof. Mechanical Engineering, Purdue University, Lafayette, Ind...........................	1893
DeBray, M...................	Professeur a l'Ecole des Ponts et Chaussees, Paris, France...	1893
Dennison, Charles S	Prof. of Drawing and Descriptive Geometry, University of Michigan, Ann Arbor, Mich....	1893
Denton, J. E.................	Prof. Exp. Mechanics and Supt. Dept. of Tests, Stevens' Institute, Hoboken, N. J.............	1893
Denton, F. W................	Prof. Min. and Civ. Eng., Michigan Mining School, Houghton, Mich...................	1894
Dixon, S. M..................	Prof. Civ. Eng., University of New Brunswick, Fredericton, N. B.............................	1894
Du Bois, A. J................	Prof. Civ. Eng., Sheffield Scientific School of Yale University, New Haven, Conn..	1894
Dudley, C. B.................	Chemist, Penn. R. R. Co., Altoona, Pa.........	1894

LIST OF MEMBERS.—Continued.

Name.	Title.	Date of Membership.
Eddy, H. T.	Prof. Mechanics, University of Minnesota, Minneapolis, Minn	1893
Edwards, L. T.	Prof. Mech. Eng., Haverford College, Haverford, Pa.	1894
Emory, F. L.	Director Indianapolis Technical High School, Indianapolis, Ind.	1894
Flather, J. J.	Prof. Mech. Eng., Purdue University, Lafayette, Ind.	1893
Fletcher, Robert	Prof. Civ. Eng. and Director Thayer School of Civ. Eng., Hanover, N. H.	1894
Foss, F. E.	Prof. of Civ. Eng., Pa. State College, State College, Pa.	1893.
Fuertes, E. A.	Director College Civ. Eng., Cornell University, Ithaca, N. Y.	1894
Fulton, Henry	Prof. Civ. Eng., University of Colorado, Boulder, Col.	1894
Galbraith, J.	Prof. of Engineering, School of Practical Science, Toronto, Canada	1893.
Giesecke, F. E.	Prof. of Drawing, Texas Agricultural and Mechanical College, College Station, Texas.	1893.
Geer, H. G.	Instr. Mech. Eng., Johns Hopkins University, Baltimore, Md.	1894
Goodman, John	Prof. of Eng., Yorkshire College, Victoria University, Leeds, England	1893.
Goss, W. F. M.	Prof. Exp. Eng., Purdue University, Lafayette, Ind.	1893.
Grover, N. C.	Asst. in Civ. Eng., Maine State College, Orono, Maine	1894
Harris, E. G.	Prof. Civ. Eng., Missouri School of Mines, Rolla, Mo.	1894
Hele-Shaw, H. L.	M. Inst. C. E. Harrison, Prof. of Engineering, University College, Liverpool, England	1894
Hering, H. S.	Assoc. in Elect. Eng., Johns Hopkins University, Baltimore, Md.	1894
Hill, J. E.	In charge Civ. Eng. Dept., Brown University, Providence, R. I.	1894
Houg, Wm. R.	Prof. Civ. Eng., University of Minnesota, Minneapolis, Minn.	1893

LIST OF MEMBERS.—CONTINUED.

NAME.	TITLE.	DATE OF MEMBERSHIP.
Hofman, H. O.	Assoc. Prof. Mining and Metallurgy, Mass. Inst. Tech., Boston, Mass.	1894
Hollis, Ira W.	Prof. Eng., Harvard University, Cambridge, Mass.	1894
Hood, Ozni Porter.	Prof. of Mechs. and Eng., Supt. of Workshops, Kan. State Agr. Col., Manhattan, Kan.	1893
Hoskins, L. M.	Prof. of Applied Mechs., Leland Stanford Junior University, Palo Alto, Cal.	1893
Hume, Alfred.	Prof. Mathematics, University of Mississippi, University P. O., Miss.	1894
Humphrey, David C.	Prof. Applied Mathematics, Washington and Lee University, Lexington, Va.	1893
Hutton, F. R.	Prof. Mech. Eng., School of Mines, Columbia College, New York City, N. Y.	1894
Jackson, D. C.	Prof. Elec. Eng., University of Wisconsin, Madison, Wis.	1893
Jackson, J. P.	Prof. Elec. Eng., Pa. State College, State College, Pa.	1894
Jacobus, D. S.	Asst. Prof. Experimental Mechs., Stevens Institute, Hoboken, N. J.	1893
Jacoby, H. S.	Assoc. Prof. Civ. Eng., Cornell University, Ithaca, N. Y.	1894
Johnson, J. B.	Prof. Civ. Eng., Washington University, St. Louis, Mo.	1893
Jones, F. R.	Prof. of Machine Design, University of Wisconsin, Madison, Wis.	1893
Keene, E. C.	Prof. Mech., Agr. College, North Dakota, Fargo, N. Dakota.	1894
Kinealy, J. H.	Prof. Mech. Eng., Washington University, St. Louis, Mo.	1893
Kimball, R. G.	Prof. Applied Mathematics, Polytechnic Institute, Brooklyn, N. Y.	1894
King, C. I.	Prof. Mech. Prac., University of Wisconsin, Madison, Wis.	1893
Kingsbury, Albert.	Prof. Mech. Eng., N. H. College Agr. and Mech. Arts, Durham, N. H.	1893
Krupsky, Alexander.	Professor of the Technological Institute at St. Petersburg, Russia.	1893
Landreth, Olin H.	Prof. Civ. Eng., Union College, Schenectady, N. Y.	1893

LIST OF MEMBERS.—Continued.

Name.	Title.	Date of Membership.
Lanza, Gaetano........	Prof. Applied Mech. Mass. Inst. Tech., Boston, Mass........	1893
Lyon, J. P......	Asst in Civ. Eng., Mass. Inst. Tech., Boston, Mass..	1894
McClanahan, Thomas S......	City Eng. Monmouth, Ill., Instr. Surveying and Eng. Monmouth College....................	1894
McColl, J. R...	Supt. Mech. Dept. University of Tennessee, Knoxville, Tenn...........................	1894
Magruder, William P........	Adj. Prof. Mech. Eng., Vanderbilt University, Nashville, Tenn....................	1893
Marburg, Edgar.....	Prof. Civ. Eng., University of Pa., Philadelphia, Pa....................................	1894
Marston, Anson.......	Prof. Civ. Eng., Iowa State College of Agriculture, Ames, Iowa...	1894
Marvin, Frank O.............	Prof. Civ. Eng., University of Kansas, Lawrence, Kansas...........................	1893
Marx, C. D............	Prof. Civ. Eng., Leland Stanford University, Palo Alto, Cal...............	1893
Marx, C. W..................	Prof. Mech. Eng., University of Missouri, Columbia, Mo..	1894
Mather, T. M............	Instr. Mech. Eng., Sheffield Scientific School of Yale University, New Haven, Conn..........	1894
Merriman, Mansfield.........	Prof. Civ. Eng., Lehigh University, South Bethlehem, Pa.....	1893
Munroe, Henry L.............	Prof. of Geodesy, Columbia College School of Mines, New York City.....................	1893
Murphy, E. C.................	Asst. Prof. Civ. Eng., University of Kansas, Lawrence, Kansas.	1894
Ordway, J. M.................	Prof. Applied Chem. and Act. Prof. of Civ. Eng., Tulane University, New Orleans, La...... ...	1894
Ostrander, J. E..	Prof. Civ. Eng. and Mech. Arts, University of Idaho, Moscow, Id..........................	1894
Owens, R. B.................	Prof. Elec. Eng., University of Nebraska, Lincoln, Neb.....	1894
Peabody, C. H................	Prof. Min. Eng. and Naval Arch., Mass. Inst. Tech., Boston, Mass.....................	1894
Phillips, A. E..........	Prof. Civ. Eng., Purdue University, Lafayette, Ind.....	1894

LIST OF MEMBERS.—Continued.

Name.	Title.	Date of Membership.
Porter, J. Madison	Prof. Civ. Eng., Lafayette College, Easton, Pa.	1893
Porter, Dwight	Assoc. Prof. Hydrc. Eng., Mass. Inst. Tech., Boston, Mass.	1893
Randolph, L. S.	Prof Mech. Eng., Va. Agr. and Mech. College, Blacksburg, Va.	1894
Raymond, William G.	Prof. Geodesy and Road Eng., Rensselaer Polyt. Inst., Troy, N. Y.	1893
Reber, Louis E.	Prof. Mech. Eng., Pa. State College, State College, Pa.	1893
Richter, A. W.	Asst. Prof. Steam Eng., University of Wisconsin, Madison, Wis.	1894
Ricker, N. C.	Dean College of Eng. and Prof. Arch., University of Illinois, Urbana, Ill.	1894
Ricketts, Palmer C.	Director Rensselaer Polyt. Inst., Troy, N. Y.	1893
Ripper, William	Principal and Prof. of Eng., The Technical School, Sheffield, Eng.	1893
Ritter, C. Wilhelm	Prof. of Civ. Eng., Federal Swiss Polyt., Zurich, Switzerland.	1893
Robbins, A. G.	Instr. Highway Eng., Mass. Inst. Tech., Boston, Mass.	1894
Robinson, F. H.	Prof. Civ. Eng., Delaware College, Newark, Del	1894
Robinson, S. W.	Prof. Mech. Eng., Ohio State University, Columbus, Ohio.	1893
Sackett, Robert L.	Prof. of Applied Mathematics and Astronomy, Earlham College, Richmond, Ind.	1893
Sherman, Charles Winslow	Instr. in Civ. Eng., Cornell University, Ithaca, N. Y.	1893
Sholl, Jacob M.	Prof. Mech. Eng., Agricultural College of Utah, Logan, Utah.	1893
Silliman, J. M.	Prof. Min. Eng. and Graphics, Lafayette College, Easton, Pa.	1894
Smith, H. S. S.	Asst. Prof. of Civ. Eng., Princeton College, Princeton, N. J.	1894
Solberg, H. C.	Prof. Mech. Eng., South Dakota Agr. and Mech. College, Brookings, South Dakota.	1894
Spalding, Fred R.	Asst. Prof. Civ. Eng., Cornell University, Ithaca, N. Y.	1893

LIST OF MEMBERS.—Continued.

Name.	Title.	Date of Membership.
Spangler, H. W.	Prof. Mech. Eng., University of Pa., Philadelphia, Pa.	1893
Staley, Cady	President Case School of Applied Science, Cleveland, Ohio.	1894
Stanwood, J. H.	Instr. Civ. Eng., Mass. Inst. Tech., Boston, Mass.	1894
Stanwood, J. B.	Director Tech. School, Cincinnati, Ohio.	1894
Stewart, C. B.	Prof. Civ. Eng., Golden, Colorado	1894
Stout, O. V. P.	Adj. Prof. Civ. Eng., University of Nebraska, Lincoln, Neb.	1894
Stratton, S. W.	University of Chicago, Chicago, Ill.	1893
Swain, George F.	Prof. of Civ. Eng., Mass. Inst. Tech., Boston, Mass.	1893
Talbot, A. N.	Prof. Municipal Eng., University of Illinois, Champaign, Ill.	1893
Taylor, W. D.	Prof. Physics and Eng , Louisiana State University, Baton Rouge, La.	1894
Thornburg, C. L.	Adj. Prof. Eng. and Astronomy, Vanderbilt University, Nashville, Tenn.	1894
Thurston, Robert H.	Director Sibley College, Cornell University, Ithaca, N. Y.	1893
Timmerman, A. H.	Prof. of Physics, State School of Mines, Rolla, Mo.	1894
Turneaure, F. E.	Prof. Bridge and Hydraulic Eng., University of Wisconsin, Madison, Wis.	1894
Unwin, Wm. Cawthorne	Prof. of Eng. at the Central Institution of the City and Guilds of London Inst., London, Eng.	1893
Voit, Ernst	Muuich, Germany.	1894
Waddell, J. A. L.	Cons. Bridge Eng., Kansas City, Mo.	1893
Wagner, J. R.	Gen. Sci. Asst. to Eckley B. Coxe, Supt. Testing Laboratories, etc , Freeland, Pa.	1894
Weihe, Frederick August	Mech. Eng., Newark, Del.	1893
Whitney, N. O.	Prof. Ry. Eng., University of Wisconsin, Madison, Wis.	1893

LIST OF MEMBERS.—Continued.

Name.	Title.	Date of Membership.
Wilcox, R. M......	Instr. Clv. Eng., Lehigh University, Bethlehem, Pa..................................	1894
Williams, S. M............. .	Prof. Civ. Eng., Cornell University, Ithaca, N Y. (Mt. Vernon, Ia.).........	1893
Wilmore, John J..............	Prof. Mech. Eng., Alabama Polyt. Inst., Auburn, Ala.................	1894
Wood, DeVolson..............	Prof. Mech. Eng., Stevens Inst. Tech., Hoboken, N. J.....................	1893
Woodward, C. M	Dean School of Engineering, Washington University, St. Louis, Mo......................	1894
Woodward, R. S.............	Prof. Mechs., Columbia College, New York City, N. Y......	1893

Supplementary List.

Howe, Malverd A....	Prof. Civ. Eng., Rose Polyt. Inst., Terre Haute, Ind..........................	1894
Hall, Christopher W.........	Dean College of Eng., Mil. and Mech. Arts, University of Minnesota, Minneapolis, Minn..	1894
McRae, Austin Lee............	Prof. Physics, Missouri School of Mines, Rolla, Mo....	1894
Murkland, C. S..............	President N. H. College Agr. and Mech. Arts, Durham, N. H........................	1894
Marstrand, Otto Julius........	Civil Engineer, Gordon City, N. Y..............	1894
Mees, Carl Teo............	Prof. Physics and Electr. Eng., Acting President, Rose Polyt. Inst., Terre Haute, Ind......... .	1894
Rice, Arthur Louis............	Instr. Mech. Eng., Worcester Polyt. Inst., Worcester, Mass...................	1894
Tyler, Harry Walter..........	Prof. Math. and Secretary Mass. Inst. Tech., Boston, Mass,.............	1894
Thomas, Robert Gibbes.......	Prof. Math. and Eng., South Carolina Mil. Academy.	1894
Vedder, Hermann Kloch......	Prof. Math. and Civ. Eng., Mich. Agr. College, Lansing, Mich.................	1894
Cumings, H. T............ ...	Instr. Civ. Eng., Union College, Schenectady, N. Y............	1894
Fava, Francis Renatus........	Prof. Civ. Eng., Columbia University, Washington, D. C....................	1894
Crenshaw, Bolling Hall.......	Auburn, Ala.............	
Kent, William..............	Mech. Eng., Passaic, N. J.......................	

Report of the Secretary.

The Society for the Promotion of Engineering Education, organized at Chicago last year by those in attendance on the Educational Section of the World's Engineering Congress, has already passed its potential stage, and it may now be regarded as an active and efficient institution. The membership was originally composed of all who registered themselves as in attendance on the Educational Section at Chicago. It is a very remarkable fact that very nearly all of those, sixty-three in number, have qualified as members. Blank forms of application for membership were at once sent out, and these have been coming back to the secretary throughout the year, so that now he is able to present to the Society over seventy applicants for membership. It is probable that before the close of this meeting the membership will number over one hundred and fifty of the leading teachers of engineering in America. together with a number of celebrated ones in Europe.

The interest in the proceedings of this meeting, also, is by no means measured by the number of members in attendance. The secretary has received a great many letters from prominent members, expressing their regrets at their inability to be present to listen to, and to participate in, the proceedings. That there is a great work for this Society to do, is patent to the entire engineering profession, both at home and abroad. Engineering education in America is still in its formative stage, and we are entirely unfettered by precedents and prejudices. With faithful effort, cordial co-operation, and continued perseverance, we now have the opportunity of ultimately developing here the "Ideal Engineering Education."

It is gratifying. also, to know that the Society has come through its first year free from debt, and has succeeded in publishing all the papers and discussions of the Educational Section of the World's Engineering Congress, in credible form, without any outside support, and without any assessments on its members. There were five hundred copies of this volume, some two hundred and fifty of which are still in hand. It is not likely, however, that the small sum of two dollars annual dues will suffice to continue these publications in good form. If this sum could be raised to three dollars, it would probably suffice to pay the bills for printing and postage, that being the only expenses of the Society at present.

If the Society prefers it, the papers could be printed in advance and supplied to the members before the meetings, at a small expense. This would require sending in the manuscript from four to six weeks before the time of meeting.

The secretary desires to call the attention of the Society to the advantage of meeting in rotation with the various national engineering societies, as well as with the American Association for the Advancement of Science. By this means we cultivate acquaintance with our practicing professional friends, and both, we and they, are likely to benefit from this more intimate association. It is very probable that such an arrangement could be made.

Your secretary also recommends increasing the membership of the council to at least thirty, and that a reasonable amount of "rotation in office" be practiced. By this means we are likely to keep out of circuitous ruts and to enlist a greater number of persons by passing the responsibilities around.

Respectfully submitted,

J. B. JOHNSON,
Secretary.

Report of the Treasurer.

I beg to report that the income and expenditures of the Society for the past year, as far as reported and paid by the treasurer, have been as follows:

INCOME FOR YEAR 1893-94.

Initiation fees and dues of members, sixty-three at $5.00......	$315 00
Books sold..	221 02
Applicants for membership paid in advance on dues for next year.....	5 50
Total...	$542 02

(2)

EXPENDITURES.

For printing of proceedings and preparing the volumes for the
mails.. .... $387 45
Printing, postage, etc.... ....   140 76

$528 21
Balance in the treasury............ .... 13 81

$542 02

The books of the treasurer will show a total income of $549.02.
The difference between this sum and the amount given as the
income of the Society at the beginning of this report is $7.00, which is
made up of $5.50 sent by Prof. Hofman as dues for the coming year,
and returned to him, therefore also appearing as expenses of the
Society; also three times fifty cents from three members, or $1.50 for
extra cloth binding. This extra charge having been abandoned, the
money was returned to the senders.

It will, therefore, be seen that the treasurer, at the end of the year,
has a balance in favor of the Society of $13.81.

Respectfully submitted,

STORM BULL,
Treasurer.

August 21, 1894.

This is to certify that we have this day examined the preceding
report and the accompanying accounts, and find all correct.

IRA O. BAKER,
F. O. MARVIN,
Auditing Committee.

PAPERS

READ AT THE BROOKLYN MEETING

ON

ENGINEERING EDUCATION.

NOTE.--The discussions at the meeting on the following papers were taken down by a stenographer and mislaid before they were transcribed. The publication of this volume has been delayed in the hope that they could be recovered. In this the editorial committee has been disappointed.

———————

This first address to this Society offers an opportunity for the presiding officer to write a book upon technical education, but having compassion upon my audience, I will be brief.

We meet under auspices favorable and peculiar. Favorable because the antagonisms of the past between classical education and scientific education have passed away. There has been a surrender or a truce. Engineering education and abstract scientific education have become closely allied, and technical schools and scientific courses are no longer suppliants for favor at the hand of the traditional educator. Meager preparation and narrow scientific courses of three years, barely tolerated, have expanded into the technical school, with its four or five years course, demanding high preparation, well manned with instructors and well endowed with its millions. For money it now appeals not in vain. Scientific education has conquered the predjudices of the past and with a broader field of view there is no reason why all learning, knowledge and education should not, under a common bond, live and flourish in peace.

21

Such men as Sheffield, Stevens, Drexell, Parker, Cornell, Stanford and others, have by their munificence crowned their names with honor. Neither should such schools be overlooked which, without munificence, have shown great faith by their works, of which the Rensselaer Polytechnic, of Troy, is an admirable type. And yet, notwithstanding great liberality, most schools feel their relative poverty. The work to be done grows in magnitude, and it is only by wisely directed efforts and faithful and frequently overtaxed men in the corps of instructors that grand results have already been achieved.

Our constitution says: "Members of this Society shall be those who occupy, or have occupied, responsible positions in the work of enginering instruction, together with such other persons as may be reccommended by the council." Whether the word male was intentionally omitted from this article I am not advised, but in these days of the great enlargement of the educational field for women, it looks like a stroke of wisdom to so frame the organic law that the admission of women to the Society might be possible without calling a constitutional convention and entering into political strife over such a minor point. Who would not feel that this Society would be honored by enrolling among its members that woman who, when her husband's health was being undermined, studied works on enginering and the plans and specifications of the structure and became the head, hands and feet of him who was the official engineer of the East River bridge, Roebling?

Or of that woman who designed for the Columbian
Exposition, at Chicago, a building to commemorate
the work of women of the nineteenth century, which,
for good taste and neat simplicity, gave a feeling of rest
and repose unequaled by any.

But this meeting is peculiar in this sense, that it is
the first meeting of a society for the promotion of engineering education in a broad sense in this country, if
not in the world. It is the outgrowth of a society of
instructors of mechanical engineering and of the enthusiasm manifested in the educational section of the
world's engineering congress at Chicago. One of its first
fruits has been the publication, under the able secre-
taryship of Professor J. B. Johnson, of the papers read
at Chicago under the title of "Engineering Education."
This book has already been of service to educators.

In less than forty years about one hundred professional engineering schools, including special courses in
universities, having come into existence in this country, graduating some twelve hundred annually. Besides
these, many other schools include some instruction in
the mechanic arts and engineering subjects. This
growth, spontaneous in its character, without a central
head or mutual conference, furnishes a sufficient reason
for the existence of this Society. If its efforts are
properly directed, it may make of all these schools a
kind of university, in which, though widely separated,
there may exist a bond of unity for accomplishing the
best results in this line of education; in which there
may be "unity in variety," as there will be "variety in

unity." Some things such a society cannot do, nor would it be best if it had the power. It cannot determine arbitrarily the proper number of schools nor the number of graduates. These will follow the law of supply and demand, and can no more be fixed beforehand than can the political economist determine the proper number of lawyers, physicians, ministers, politicians or any element of the body politic. It may be safely asserted that there is not a demand in the country for one thousand or more new professional engineers annually, but the fact that that number find useful employment, and that these schools are more and more crowded with applicants, shows that this kind of education is growing in popularity. But all of the graduates do not follow the profession of their school for life's work. These schools have opened new lines of work, and raised the standard of others, as well as enlarged the field of engineering. Graduates are found in many, if not in all of the other learned professions, and in many departments of business. It therefore becomes doubly important that this education should be conducted on correct principles; that while the subjects for study are those of engineering, they should be so handled as to develop the mental powers. A cast iron course cannot be made, much less enforced. There must be a certain amount of flexibility; certainly, of personality.

We shall consider three topics: First, what should be taught in engineering schools? Second, how taught? Third, who should teach?

Library
N. C. State College

On these points we will speak of such underlying principles as are common to all kinds of engineering education—civil, mechanical, electrical, mining, military, naval, sanitary, and others, if there be such. A young aspirant asked an eminent engineer what he should study to be of most benefit to him in practice. He replied: "Chinese." While there is no danger of the schools abandoning their well chosen technics, yet one may accept the spirit of the answer and despise not the learning of anything which comes in his way.

Our hopes of the future are built on the experiences of the past, but from the same history may be drawn different lessons, depending upon the bias or depth of thought of the reader. We will attempt to draw one. About twenty-three centuries ago Socrates conversed with a few young Athenians, not on amassing wealth, nor of securing official station, nor of gaining power over men, but upon the relations of the universe, of virtue, of immorality, of God. It was a small beginning, a private school, but out of it came Plato, and through his school Aristotle, a giant intellect, who wrote upon nearly every subject of human thought then known. He was not backed by a church nor a state, nor by an army, but his philosophy held sway over men for more than eighteen centuries. This domination of intellectual power by one man is a sublime fact in history. And yet his philosophy of physical science contained such weak and glaring errors that it required only a simple experiment to show their fallacy but during all that long period no one was thoughtful enough, or rather brave enough, to expose them. Of these one example will serve our purpose.

Aristotle taught that bodies of unequal weight would fall in equal times through space proprotional to their weights. Or, more simply, that a two pound weight would fall twice as fast as a one pound weight. At last a man arose who challenged the truth of this assertion, Galileo claimed that the philosophy was false, and in 1590, only 304 years ago, challenged his opponents to an experimental test at the tower of Pisa. On one side was the tradition of centuries, the upholders of Aristotelian philosophy, professors in the university jealous of the popularity and rising reputation of the young philosopher. students who clung to the popular side and the superstitious who would not, if they could believe in anything new; on the other side was Galileo with two balls in his hand ready for the experiment, full of faith and hope for the truth, and about him a throng who "cared for none of those things." It was a momentous occasion, two theories vital in the interest of truth were on trial. One or the other must go down. The balls falling from Galileo's hand striking the earth at the same moment sounded the death knell of the Aristotelian dogma. Subsequent history exalts the significance of that hour. Aristotelian physics was an utter failure, for it drew no laws from facts or in conformity therewith, and hence was powerless to explain nature. If a fact did not conform to theory so much the worse for the fact. Since that hour facts have been the weapon for battering down false theories, however venerable or deeply rooted. A fact is a truth, and a truth is the evolution of the ages. It has, figuratively speaking, been deified.

It would be natural to suppose that the experiment of Galileo, so simple and conclusive, would have been universally accepted, but it was not. With some it is easier to cavil than to reason, and to reason than to believe. All ages have scoffers whose chief argument is a sneer, and that period was not an exception. Galileo's teachings brought upon himself bitter persecutions; but he had planted a seed of truth which took root and flourished. The spirit of true scientific research had asserted itself. The history of such an age is a lesson to all subsequent ages. It has taught that truth is not to be sought in dogma or tradition but by honest, careful study of facts whether exhibited by nature or determined by experiment; it has also taught the worthlessness of mere speculation. Still speculations are profitable, since they raise questions for consideration and often lead to beneficial results; but only those are valuable whose results conform to facts. All theories are at first speculations. Newton held to his conception of the law of gravitation for fifteen years before he proved it. As light creeps on gradually from night to full day, so it is with the mind as it searches after truth. Facts alone will not interpret nature—theory and law are essential. Physical law is the golden chain which binds together a series of facts into one common principle. It is the creation of the intellect—a human product. There was a time when the formu-

lated laws of nature were considered as the very
thoughts of the creator, but precision in measurements
and care in experiments have rendered it highly probable that no formula represents exactly operations in
nature. The law of gases as first formulated, are
now known to be imperfect, and no perfect formula
has been established for them. Even the Newtonian
law of gravitation is not known to be exact for infinitesimal distances nor if its action is instantaneous.

Galileo was the first dynamic engineer, not in the
sense that he constructed any engineering works but
in the sense that he made dynamic engineering possible. He was the first scientific writer on the strength
of beams. Problems involving the motion of masses
could not be solved before his day, for the laws governing them were unknown, but he laid the foundation
for such solutions by discovering the necessary laws
which were formulated and known as Newton's three
laws of motion.

Strange as it seems engineering practice involving
the more refined and recondite theories of the unknown
and molecular actions have with a leap sprung into the
foremost ranks. Electrical engineering is as closely
allied to theory as any other line of engineering.

Briefly, then, whatever subject be studied, whatever exercise be performed, whether in the lecture room,
laboratory or shop, let there be research, an inquiring
into facts, theory, law; these three, but the greatest of
these is *law*.

If education be planned for merely utilitarian ends,

if it is sought to transmute every formula, whether of mathematics, physics, chemistry or literature, into gold, education would not only degenerate, but lose its power as an elevator of the intellect. Love of knowledge for its own sake is the highest motive for study. But the strife for place and early returns, and the comparative poverty of many who seek professional life, tempt if not force them to take a lower standard. The very keynote of imparting knowledge is the use of the free inquiry instead of the dogmatic method. The student should be led to discover truth as if it had not been discovered a thousand times before. The value of this method has been recognized and resulted in the establishment of many laboratories and shops, in all of which the student may make the tests, experiments, and even learn handicraft. For a long time experimental research was considered a pastime for a few, but so many benefits have resulted to mankind from such investigations securing commercial wealth and its attendant comforts and luxuries, that scientific education has come to be considered of national importance.

Apparatus cannot be a substitute for a live teacher but is of great aid to him. Models for illustration need not, as a rule, be costly; all that is necessary being such as will enable the learner to gain a correct idea of the relation of parts. For this purpose, sometimes a model costing a few dollars will be as serviceable as one costing hundreds. But for original research in some cases the mechanism can not be too accurate. A text book in which the matter is well selected, well stated, well digested, well arranged and well printed, is one of the best pieces of apparatus for the learner. He will feel more confidence in the statement of an author than in notes taken from the utterances of a lecturer, and if in after years he has occasion to use the results sought, he will sooner turn to the book than to his lecture notes. Indeed a text book may be considered as a course of lectures printed after thorough revision. Lectures, however, should not be ignored or suppressed, but should supplement the texts, where books of reference, periodcals, and current literature may be brought to the attention of the student.

Thirty or forty years ago there was a dearth of text books in certain technics, but now there is such a multiplicity of them and such a desire to cover a large field, that there is a temptation to "cram." It is not the amount and variety of the viands put before the epicurean, or the amount taken into the system, that makes the bone, sinew and physical vigor, but that which is digested and assimilated. All are learners; and graduates, instructors, and amateurs will find decided advantage in becoming members of one or more of the scientific professional societies. Personal experiences, discussions, and interchange of views are of great profit to all concerned.

It is not the object of the school to make engineers, but to educate them. Engineers are born, not made. The man chooses his profession, the professor instructs; but no amount of school training will supply brains, but it may enable one to use even limited powers more efficiently. These ideas will give tone and character to the instruction.

All physical laws are founded on statistics. Statistical and historical facts underlie laws; their relations are that of ruler and subject. How much of history and statistics shall be studied depends so much upon circumstances that no rule can be given; but the man is generally better equipped for the present and future problems in life who well understands its past. Knowledge is not complete without history and statistics. To pursue either exclusively is not education. The acquiring of any number of facts is not education. Much less is that education which merely acquaints one with the fact that certain formulas are in certain books. The possession of a library, however valuable, does not make one learned.

A distinction may be made between the work of a beginner and that of an investigator. The former accepts without question those laws which by tests and experience have been considered as fixed, such as the laws of falling bodies, the laws of gravitation, etc., and then proceeds to ascertain the reason for them. The investigator may question the accuracy of any law and proceed by more careful measurements to determine anew the data; or he may proceed by the statistical method to determine undiscovered laws. Throughout history science has shaken much, but increased knowledge more.

The teacher should enforce the importance of both accuracy and precision. The distinction between these

is important.   To illustrate, I will take an example in
my own practice.   I directed an assistant to set the
transit over a certain stake and give the line.   He did
it with great precision placing the bob exactly over the
nail, leveling the plates very accurately, adjusting the
eye and object glasses very nicely.   Then I found that
he had set the instrument over the wrong stake two
rods from the correct one.   He was precise but not
accurate.   Calculations are sometimes carried to many
decimal places, giving the appearance of great precision
when from the nature of the case the data is imperfect.
The value of the degree of precision depends upon circumstances.   An error of 100,000 miles in the distance
between the earth and sun would be an error of about
one-tenth of 1 per cent., but an error of 1-100,000 of an
inch in determining the length of a wave of light
would be an error of about 50 per cent. of the correct
value.   In construction one must judge for himself
where precision is justifiable.   Precision may be used
as a quibble for dishonest purposes.   A fly was once
put in a dead man's mouth that a witness could swear
there was life in the man when he signed the will.   One
was asked if it was true that he suppressed the document.   "Why, that document was printed," he replied.
And so it was several weeks after the suppression.   His
answer was precise, but he was dishonest.   Precision
may or may not be the truth; accuracy is truth.

It is quite possible that the "what" and "how"
herein set forth appear too reasonable to be questioned
but one need not go far to find that perpetual motion-

ists, and even believers in a flat earth are not extinct. Visionary theories are made on doubtful or even speculative grounds. Quack scientists are abroad. It belongs to the school to conserve the truth, and to correct processes of thought and investigation. In doing this too much care can not be exercised in enforcing correct views of the fundamental principles of a science. This, at least as an educational element, is especially important in regard to the science of mechanics. The appropriateness of the elementary terms used and the definitions are still discussed, although supposed to have been fixed long since. Reference is not here made to the different wordings of the terms nor modifications resulting from increased knowledge, but to differences which are essential. The substitution of the term "energy" for "living force" has removed an ambiguity which was more troublesome than vital. It may be said that literally, volumes have been written upon the one term, "momentum." The erroneous view held before Galileo's day, and at one time entertained by him, that momentum is force, is still upheld by some writers not of the highest rank. The exercise of free research is not denied in this, any more than in other things, but the opinions and results of research of such men as Galileo in his later years, of Sir Isaac Newton, Thomson, Tait, and Rankine ought to have more weight than those of a novice. There are no differences, so far as I am aware, among the best writers, of which Thomson and Tait, in their Natural Philosophy, may be taken as a modern type.

3

The "how" consists largely in teaching the learner how to think, how to determine facts, how to test theories; and not, as the story runs in regard to the theological professor when asked a question—"Young man, I wish you to understand you are not here to think, but to absorb."

Who shall teach? The instructor in engineering education should be one who loves to teach, who has a love for the truth and a fondness for his specialty, whose integrity is above suspicion, who has the necessary qualification and who so assimilates knowledge that he will give something of himself to those he instructs. That instructor is fortunate who possesses personal magnetism, and when combined with a strong, pure character he exerts an unconscious power for good over his students.

It is not necessary that the instructor should be an expert in the shop, the mine, or the field; but a practical knowledge with the arts involved in any, or all these, better fits him for his duties. As a rule it is not desirable that a teacher of general engineering should be an expert in handicraft or be eminent in a specialty; for, however much he might desire to avoid it, he would be liable, unconsciously, to give undue prominence to his specialty to the subordination of other subjects equally, if not more, important. The constructor of a model Jerusalem became convinced that he had made a more accurate representation of the city from reports of surveys and photographs than if he had visited it. An eye witness gives prominence to such things as impress him most strongly, but the student of facts drawn from a variety of sources is comparatively free from such impressions. Similarly with the instructor.

At first sight it seems paradoxical that some of the best instructors of technical education have come from those who have had little experience in the arts. An instructor may be a better teacher than practitioner. Inventive genius, natural abillity in the art of construction, a thirst for technical knowledge, and a deternation to accomplish certain results are all important elements. Inventions have been made by those whose studies and pursuits were foreign to the subject A college bred man and an artist invents the telegraph, another unacquainted with the trade invents a shoe machine, while another, feeling the want of early advantages, gives his millions to form schools for others.

Instructors should not be triflers. Sufficient time has been wasted over the question "What would be the result of an irresistible force meeting an immovable body?" to gain a knowledge of Newton's three laws of motion. He should not be enigmatical. Some profound writers have an abillity of stating principles in such language as to require patient, prolonged, profound study to determine their hidden meaning. Such writing, though often of untold value to the science should be translated and interpreted by the instructor before presenting the subject to the beginner. He should have a scholarly spirit. A love of knowledge for its own sake and not for commercial purposes is the inspiring sentiment of the true student. He who discovers truth in this spirit is repaid by the joy experienced. If this love of learning be not possessed by the instructor he can not impart it to others. Remove it and study becomes mechanical and a burden, a kind of tread-mill work without heart and years of labor may be wasted.

Learning merely for the purpose of appearing to be learned, or merely taken in to be given out does not promote culture, but may give a certain externality. Whitewash, thick or thin, is only a temporary covering. If one appears to exhaust himself with every effort he can not exert a lasting impression. The living spring only is exhaustless. Muddy water may be mistaken for deep water, but the illusion is dispelled by the measuring rod.

Great men receive knowledge from every source, even from a student; and he who can, without injury to himself, say "I do not know," is truly great. A professor once made this noble remark to a class, "I care more what you will think of me twenty-five years hence when you are fighting life's battles, than what you think of me now." Let such far sighted wisdom be the sentiment of this Society.

This Society is to exert an influence, and it is hoped a lasting influence, on engineering education. Let the foundation be well laid to serve, not only the purposes of to-day, but of the future. Let it be enduring. The possibilities of the future are great with hope, and the educated engineer is to be one of the important factors in the development of the resources of the world.

Already the lives of individuals and the fortunes of men are at his disposal.

"Time is money" in a more utilitarian sense than ever before. Two merchants from the same town buy a stock of goods at the same place; one ships his by a cheap but slow water route, the other by express at a higher rate. The latter receives his goods promptly and sells the largest part in a quick market; when the former receives his the market has been supplied and he may sell at a sacrifice. Thus is learned the fact that time saved is money gained. This illustration shows in part why so many canals have been dried up by the locomotive.

The real question in this, as in all legitimate business, is not how shall waste be reduced to the smallest amount, but rather how shall the highest efficiency be secured when all the elements involved are taken into account? For instance, the least waste on a railroad might be secured by letting the plant stand idle, for then there would be no expense for labor, fuel, wear and tear, or accidents, and the detrioration might be less than when in operation; but on the other hand, there would be no earnings. Waste accompanies profit.

The naval battles of the present are fought by the engineer. The Yankee cheese-box on a raft revolutionizes systems of warfare. It is now the battle of the metals. The success of land battles may yet depend more upon the skill of the engineer than upon the number slain. Rapid communication and construction have done much in alleviating the distress of those who carry arms in the most horrible method of settling disputes.

But the greatest monuments to the engineer are found in the arts of peace. The combination of the engineer, the inventor, and the capitalist in the making of canals and railroads, the extraction of metals, transportation by steam, the application of electricity, agricultural implements, sanitary improvements, and of numerous labor saving machines, are modifying the habits and customs of society, and producing a greater revolution than the armies of a conquerer. The labor saving machine is also labor making. It not only opens a new line of manufacture, but furnishes new fields of labor for those displaced. It forces a redistribution of labor. It may increase the number of mechanics, shippers, salesmen, clerks, officials and promoters of new enterprises, or it may increase some of these and diminish others; but so complex are these changes that no lynx-eyed observer can trace their bewildering paths. These energies give increased power to those who direct them. In the hands of the benificent they increase comfort and blessing: in the hands of the selfish and designing, misery and degradation. Education elevates.

It is the province of the engineer to harness the forces of nature and make them serve the interests of man. With increased facilities for storing and transmitting energy, wind, waterfalls, and tide will be more useful in supplying power, light, and heat where desired. Thus the scientific method is working itself out in various directions, and it is the province of this Society to consider the means of promoting that education which shall be best for those who are to labor in this sphere of life's work.

COMMON REQUIREMENTS FOR ADMISSION TO ENGINEERING COURSES.

By FRANK OLIN MARVIN,

Professor of Civil Engineering, University of Kansas, Lawrence Kansas.

As an introduction to this subject, the writer has thought best to briefly outline the present situation in America, as regards the requirements for admission to engineering courses. The facts presented have been taken from the latest catalogues ('93 and '94) of seventy-seven colleges and schools, representing all phases of engineering education, from the advanced work of the higher schools to the meagre courses of some of our schools of Agriculture and the Mechanic Arts. The character of the courses generally indicates the character of the standards of admission, but not always. A few schools have high requirements with a low grade course, while the reverse is true of a few others. For the purposes of this paper, the seventy-seven schools will be grouped in two classes only. Class A consists of those of average higher standing before the public, and as indicated by the admission requirements and nature of the courses offered, and numbering forty-three. Class B numbers thirty-four, and contains schools of lower standing and less recognition at large.

Following is a tabular view of the subjects used as requirements, and the number of schools requiring each:

SUBJECTS REQUIRED FOR ADMISSION TO ENGINEERING COURSES IN 77 COLLEGES, AND SHOWING THE NUMBER REQUIRING EACH.

SUBJECT.	CLASS A. 43 Schools.		CLASS B. 34 Schools.		TOTALS. 77 Schools.		
	Requiring.	Not Requiring.	Requiring.	Not Requiring.	Requiring.	Not Requiring.	
PENMANSHIP,			5		5	72	
READING			6		6	71	
SPELLING	4	39	10	24	14	63	
GEOGRAPHY	19	24	23	11	42	45	
MATHEMATICS:							
Arithmetic	34	9	28	16	62	15	
Algebra, Element'y	*3		15	10	18	10	*May be through quadratics.
Through Quadratics	†22		8		30		†One school counted twice.
Advanced	†15		1		16		
Complete	4				4		
Geometry less than 5 books.	2		5	23	7	23	
Plane or more	15		3		18		
Complete	26		3		29		
Trigonom'try, Plane	*7	33		34	7	67	*3 add. optionals
Complete	†3				3		†1 add. optional
Mensuration	2						⎞
Compass Surveying.	1						⎟
Desc'pt've Geom'try	1						⎬ For Post Graduate Courses
Analytical Geom'try	2						⎟
Calculus	2						⎟
Elementary Mechs.	1						⎠
DRAWING	6	37	1	33	7	70	
HISTORY:							
United States	31	12	19	15	50	27	
England	*2		2		4		*3 add.optionals
Greece and Rome	†2		†2		4		†2 add.optionals
General	††6				6		††5 add.optionals
Civil Government	‡3		3		6		‡3 add.optionals
ENGLISH:							
Grammar	29	14	26	8	55	22	
Comp. or Essay	37	6	18	16	55	22	
Rhetoric	11	32	6	28	17	60	
English Literature	5						1 add. optional
Classics—under 5	7	15	2	28	9	43	
5 or over	21		4		25		
No English				5			
SCIENCE:							
Physical Geography	*10	33	12	22	22	55	*4 add.optionals
Physics	†18	25	†8	26	26	51	†7 add.optionals
Chemistry	†††7	36	3	31	10	67	†††7 add.optionals
Physiology	‡6	37	‡6	28	12	65	‡8 add.optionals
Botany	**4	39	1	33	5	72	**8 add.optionals
Zoology							5 add.optionals

SUBJECT.	CLASS A. 43 Schools.		CLASS B. 34 Schools.		TOTALS. 77 Schools.		
	Requiring.	Not Requiring.	Requiring.	Not Requiring.	Requiring.	Not Requiring.	
Astronomy	3						
Geology							2 add. optionals
Meteorology	1						
Mineralogy							1 add. optional
Physiography							1 add. optional
GENERAL:							
Bookkeeping	2						
Com. Arithmetic							1 option
Elocution							1 option
Mental Science							1 option
Theory of Teaching							1 option
Political Economy							1 option
Etymology							1 option
FOREIGN LANGUAGE:							
Greek							1 option
Latin							2 options
Latin and [Physics or Greek]			1		1		
French	2				2		1 add. optional
German and Latin, or German or Latin	1				1		
Latin or English	1				1		
Latin or Science	1				1		
Latin or German or French	4		2		6		
French or German or English	1				1		
French or German or Mathematics	1				1		
French or German	8		2		10		
[French and German] or Latin			1		1		
[French or German] and [Latin or Eng.]		24		25		49	
[French or German] and Latin	1				1		
French and German with Latin for either	1				1		
French and German	1				1		
French or German; French or German and Latin; French and German	1				1		
Latin and Greek			1		1		Either combination
Latin and Greek and German			1		1		
Latin, Greek, French German, Spanish							1 optional, singly or any comb'n'n

Attention is especially called to the following points: Many of the published statements are more or less vague, either as a whole or in certain parts. Mathematics and Latin appear to be the most definitely presented, though Algebra or elementary Algebra are sometimes named without any limitation. The subjects of Physics, Chemistry, Physiology, Botany, English Literature, Rhetoric and history sometimes appear without any indications of the quantity or kind of work demanded. The language requirement often is measured by time expressed in years, without any qualifications as to the ground covered or the number of weekly exercises. In two or three cases where options are allowed between different languages, the phrasing is such as to leave it uncertain what combinations are permissible. On account of this vagueness, the above table may be in error in some details. In justice to those seeking admission, the catalogue statements should be both full and clear.

The study of this subject has brought out the fact that there is a great variation in requirements, not only between the lower and higher classes of schools, but also between the individual schools of the same class. This is due, perhaps, to the recent rapid growth of schools of engineering, to local conditions affecting the character and standing of the secondary schools which furnish the students for these courses, and to the fact that so many of our engineering schools have had their beginning as an offshoot from the old classical college. This variety, however, makes us, as a body of teachers, liable to the charge of not knowing what we want, of not clearly perceiving what things are essential in the secondary education that leads to the engineering course. This variation finds expression, as well, in the college courses, as regards subject matter, the sequence of subjects and the relation of one subject to the course as a whole.

The retention of the common branches, Geography, Arithmetic and formal English Grammar among requirements, by respectively 19, 34 and 29 schools of Class A, is to be noted. Where examinations are held in Physical Geography, Algebra and Geometry, and Higher English, these might well be abandoned. The same might be said of U. S. History (required by 31 schools of Class A), where examinations are held in General History or other advanced historical subjects. The mathematical requirements show the greatest uniformity. Twenty-two schools of Class A require Algebra through quadratics, 15 others require something over this amount, while but 4 require its completion. Twenty-six schools of Class A require all of Geometry, and 15 either plane Geometry or a little more, with only 2 asking for less than five books. Trigonometry is apparently above the reach of the great majority of schools at present.

In History, only that of the U. S. shows any numerical strength. In English, Grammar and an essay seem to meet the views of the larger number, with 28 schools of Class A requiring English classics from two to ten in number. In Science, Physics and

Physical Geography are found to be the requirements of about one-third of all the schools in both classes.

Note the fact that Drawing is required in but seven schools of the entire number. There are but 36 schools out of the 77 that ask a foreign language, or languages, for admission, and in 8 of these the language is optional. This makes 28 schools only that require a foreign language, of which number 19 are in Class A. Twenty four of Class A and 25 of Class B require no language.

The following will show the possibilities as to entrance in this line:

Foreign Language.	No. of Schools.
Admission with Latin only..............	12
Admission with French only.......................	20
Admission with German only....	18
At least one language required	23
At least two languages required................... ..	4
At least three languages required	1
No language required, but some language optional...	8
Total number of schools that ask foreign language work for admission............	36

French *or* German is the combination offered by 10 schools, and French *or* German *or* Latin is offered by 6 institutions. Each of the remaining schools has an unique arrangement.

To further show the variation that exists as to foreign language work, the following table is given, indicating the requirement for graduation, including the entrance requirement:

	Class A.	Class B.
Requiring no foreign language............	7	16
Requiring one foreign language...	18	12
Requiring two foreign languages.......	19	4
Requiring three foreign languages....	1	2
	45*	34

*Includes two schools having different quantities for different courses.

When it comes to a consideration of the time given to foreign language work, or to a comparison of the ground covered, the problem.becomes indeterminate, when based on catalogues. Suffice it to say, that the quantity seems to vary between the limits of one year of one language, and three and one half years of one language, or two years in each of two languages. The lowest amount of any one language required in the entire course by one of our most prominent schools, appears in the catalogue as the equivalent of one-half year of French, five times a week. Yet the writer is sure, from a personal acquaintance with the instructor, that much more is attained by the student than the time given would seem to indicate.

To summarize the situation, the schools in Class A are practically agreed on the following requirements: Arithmetic; Algebra through quadratics, or a little beyond; plane and solid Geometry; U. S. History; English Grammar; an English Essay; some English Classics; and, possibly, Physics and some one foreign language.

The schools of Class B are practically agreed on

requiring Geography, Arithmetic, Elementary Algebra, U. S. History, English Grammar, and English Composition or an Essay. The above constitute low standards of requirement, and do not represent the actual averages, though very close to them, inasmuch as some individual schools add other subjects, while others require less than the amount above stated.

This brings us to the main question, what should be the

STANDARD REQUIREMENTS.

There are a few general principles which should govern in this discussion.

First—The high schools are the natural feeders of our colleges and schools of science. The larger number of these institutions must look to them for their constituency, rather than to private schools. This applies with especial force to the State Universities, which are maintained by the people at large as a part of a state system. This is only of little less import to the colleges built on private foundations. No college can afford to ignore the public high school. On the contrary, the college should keep in touch with the public school, manifest interest in it and exert a reflex influence over it. The essentials of the life of the public school, and through it, the life of the country we love are locked within this relationship.

Second—There should be no gap between the high school and the college. The pupil can not often know whether a college education is desired or not, or can not see the way open to fulfill such a desire, until the close, or near the end of the high school course. The way ought to be clear to enter some college course when such choice is made. No stumbling block, in the shape of a year's additional preparation, should be in the road to divert the student from his purpose, or turn him toward an inferior school.

Third—It follows that the college requirements for admission should coincide with the high school requirements for graduation. It is not meant by the above that a high school classical graduate should have the way open to enter the Freshmen year of a scientific or engineering course, though he may have an aptitude for such study; only this, that the high school should prepare for the college, and that the college should accept the standard.

Fourth—It must be recognized that the high school is not primarily a fitting school. Its office is to give American youth the best training and truest culture that it is possible to furnish them by the time they are 18 years of age. Preparation for college is incidental.

Fifth—Preparation for college requires no different treatment of the pupil as to the method of instruction from that required for his high school graduation. (See the "Report of the Committee of Ten.")

Sixth—The requirements for admission to Engineering courses should be different from those required for Arts courses as to some subjects, including less foreign language and more Science and Mathematics.

Seventh—There is no proper admission requirement for four year Engineering courses of average high standing, that may not appropriately fit into one or more courses of the high school.

No disparagement of private preparatory schools is intended. They have their place and mission. But from the point of view of this paper, they are without this discussion, inasmuch as they are forced to conform to college requirements to maintain their standing and prestige.

The following subjects appear to the author to constitute the proper requirements for admission to an engineering school of high grade, one that seeks the training of professional engineers, rather than the education of surveyors and mechanics. In Mathematics, Algebra, through quadratics, ratio, proportion, progressions, the binominal formula and logarithms; plane and solid Geometry; plane Trigonometry. This would be followed in the Freshman year by higher Algebra, a rapid review of plane Trigonometry and the completion of spherical Trigonometry and Analytical Geometry. In Science, Physical Geography, Physiology, Elementary Botany, Physics and Chemistry are the subjects desired. The three last named should be taught in the high school by the laboratory method. Botany is chosen rather than Zoology, because of the greater facility with which laboratory material can be secured, and because of the more general interest in it. Physics and Chemistry should each be carried through a year with the equivalent of five exercises a week, at least one-half of the time being given to laboratory work. Entrance examinations in Botany, Physics and

Chemistry should include the presentation of certified note books, showing laboratory examination of 30 or 40 plants and flowers, 40 experiments in Physics, and 60 or more experiments in Chemistry.

In History, the high school course should include history of Greece and Rome or General History, English History, and Civil Government. U. S. History is assumed as belonging to the work of the common school.

Drawing is properly an entrance requirement, and the examination should include an inspection of the figures in the laboratory note books, and a test of the ability of the applicant to make a quick sketch, showing the degree of his mastery of the art as a means of expression. Throughout the preparatory course, Drawing should be taught as a language, and should be kept in constant use in connection with Geometry, the Sciences and History. The old method of servile copying from the flat, for the purpose of making a pretty picture, is of no value, and should find no place in the schools.

In English, the entrance examination should presuppose a most thorough training, not only in Grammar and Elementary Composition, but also in Elementary Rhetoric, and in a rapid and forcible use of the mother tongue. It should also require a considerable knowledge of the classics of English Literature. It is not sufficient that the applicant should have read a limited number. He should have an intimate acquaintance with the story, the plots, the characters and the

4

true meaning of at least ten masterpieces, chosen not only for their intrinsic worth, but also to illustrate the different periods of English Literature. Formal Rhetoric, so far as it relates to complicated questions of taste and style, and the so-called history of English Literature, as found in the usual manual, should find no place in either the high school course or the entrance examination. The ability to write good English should be determined, not only from the formal English examination, but also from the papers presented in other subjects. The combination of a passing grade in English with execrable English in other papers is no rarity in the September of every year.

In Foreign Language work, there should be required both French and German, of an amount equivalent to two years work in each, with three or four exercises per week. This might be followed in the Freshman year by a year of Language, preferably German. In case Latin should be offered for the entrance French, the latter language could be taken during the Freshman year. The entrance examinations in French and German should include the main facts of Grammar, but be largely directed to a test of the student's ability to read and translate ordinary prose at sight.

The examinations in all subjects should be designed to test the power of the candidate over the knowledge acquired, rather than the quantity of information he may have gathered.

The above requirements, taken together, make a good high school course on the one hand, and seem appropriate and wise as a preparation for engineering study on the other. The professional engineer of to-day is called to assume a place among men of power, of culture and refinement. If he is to stand well among his peers and win from them a hearty recognition, he must not only have the power that comes from an intensive study of things that are of acknowledged utility, but also the breadth of view, the sympathy with the best life about him, that can only be attained through a liberal course of study. If an engineer's school life is to close with the senior year, the college course must be given over largely to technical training. The high school course on the contrary can not be technical. Its office is to give general training and culture and, too, at a time when the strongest impressions are made. The Mathematics, the Drawing, the Physics and the Chemistry are essentials for the subsequent technical study. The Literature, the Languages and the History are equally essential for the engineer as a man among men.

DIFFICULTIES.

There are many difficulties in the way of the adoption of such a standard, and they are not to be easily overcome. To cover the ground indicated would require a four years' course with twenty exercises per week, including laboratory and other unprepared work. Many of our towns and smaller cities are unwilling to maintain a course of this length. There is among the people at large a lack of a true appreciation of the essentials of a secondary education, and of the value of

modern science and its laboratory methods as a part of
it.   There is opposition to the introduction of modern
language, especially French, into school courses.   The
money for laboratory facilities and for trained teachers
to direct their use is not always at hand.   The energies
of many schools are scattered over too many subjects
and too many courses.   There is as great a diversity
between the school courses of different towns as is
shown by college requirements.   There is a lack of an
authoritative standard.   There often exists a feeling of
antagonism between school people and college people.

REMEDIES.

The remedies must largely lie along the line of
agitation and discussion, and in this, the colleges,
through presidents and professors, must take their
part.   They must manifest interest in the schools as
institutions belonging to and for the people, in the face
of accusations of selfish interest.   They must encourage and help them by personal contact, through visita-
tion,   by lectures, addresses   and publications,   by
association with teachers at conventions.   The colleges
have a further duty to perform in bettering the training
of teachers for the secondary schools, and in pointing
out better methods of work, particularly for the modern
subjects of science.   Conference and mutual agreement between colleges will do much toward simplifying
the problem.   Their united action will eliminate points
of difference between themselves and bring order out
disorder, and their conclusions will have weight.   The
action of the New England Association of Colleges and

Academies, concerning the English and Latin requirements, is felt all over our land. There is no summing up the influence that will be exerted by the "Report of the Committee of Ten."

WHAT CAN BE DONE NOW.

The writer believes it practicable for the great majority of the Engineering Schools of Class A, and a few of Class B, to agree on requirements that would represent the work of a three years' course. This course would include Algebra, thorough quadratics; plane and solid Geometry; Physical Geography; Physiology; Physics and Chemistry with laboratory work; History of Greece and Rome, or History of England; Civil Government; Drawing; English Composition and Classics, a good stiff course through the three years; and two years of French or German, four times each week. The loss from the preceding would be something from History, English and Drawing; all of advanced Algebra, plane Trigonometry, Botany and one Language. The securing of laboratory work in Physics and Chemistry would be the greatest difficulty. But this ought to be found in every high school course for the sake of every boy or girl, irrespective of the question of a college education. Much can be done with simple and inexpensive appliances, and it, perhaps, lies with the college to point out how to attain the maximum results with minimum means. One thing can be demanded of our applicants at once, and that is better training in English. The improvement desired does not consist of more knowledge of the rules

of Grammar and the meaning of rhetorical terms, or
the ability to parse and to analyze complicated sen-
tences, but rather implies an increased facility in the
use of good English, a larger power over it as a means
of expressing knowledge, and a closer acquaintance
with some of the best literature. Some of the schools
do the work now; all can do it and ought to, for their
own sake.

There is one phase of this subject that has not
been touched upon, viz: that relating to the majority
of the schools that have been placed in Class B. No
discourtesy is intended by the apparent neglect. The
limits of this paper preclude any proper discussion of
their relations to schools of higher grade or to engineering education. Situated as many of them are,
they are forced to maintain low standards of admission,
whatever may be their aims or desires. The writer
believes that they are doing a noble work in a field
peculiarly their own, and that there is a mission for
them which will exist for years to come. They are
differentiated from the higher schools by the very
nature of this mission, and this fact ought to be recognized in the degrees given on graduation.

In the movement that is now on foot, looking
toward changes in the scheme of primary and secondary
education, it behooves teachers of Engineering to take
a hand. For they are equally interested with the professors of Arts courses, and, let us hope, as earnest and
patriotic, and as desirous of the best education possible
for the American boy or girl. So the author expresses
a hope that the topic which forms the caption of this

paper may receive the attention and discussion that it merits, and suggests that this Society should appoint a committee to consider the entire question, with instructions to report at the next meeting. This committee should be large enough to be representative, so that its conclusions, approved by the action of the Society as a whole, may have due weight and authority.

APPENDIX.

The following courses for high schools are given, not as the best ones that could be made, nor as indicating the wisest arrangement of studies, but only to show that the foregoing requirements for admission can be compassed within the high school period. They are based on the supposition of 20 exercises per week for each pupil, one-fourth of them being given to laboratory or other work requiring no previous study.

FOUR YEARS' HIGH SCHOOL COURSE.

FIRST YEAR.

Elementary Algebra... 5
English History....... 5
English, Literature and Composition 5
Physical Geography........... 3
Drawing 2
——
20

SECOND YEAR.

Plane Geometry......... 4
English, Literature and Composition.................... 4
History of Greece and Rome... 3
*Botany... 3
French or [German]........... 4
Drawing...... 2
——
20

THIRD YEAR.

Advanced Algebra............. 2
Solid Geometry.... .....`···.. 2
English, Literature and Composition 3
*Physics 4
French or [German]. 3
German or [French]........... 4
Drawing 2
——
20

*Laboratory practice Saturday mornings.

FOURTH YEAR.

Advanced Algebra............. 2
Plane Trigonometry......... . 3
Civil Government, one-half yr $\big\}$. 4
Physiology, one-half year
*Chemistry..... 4
English, Literature and Rhetoric... 3
German or [French]...... ... 3
Drawing.................. ... 1
——
20

The apportionment between the different lines of work in the above is as follows, viz: Mathematics, 22½ per cent.; Science, 20 per cent.; English, 18¼ per cent.; French and German, 17½ per cent.; History, 12½ per cent.; Drawing, 8¾ per cent. Latin could be substituted for either French or German.

THREE YEARS' HIGH SCHOOL COURSE.

FIRST YEAR.

Elementary Algebra........	5
History of England, or Greece and Rome........	5
English, Literature and Composition	5
Physical Geography	3
Drawing	2
	20

SECOND YEAR.

Plane Geometry.	4
English, Literature and Composition	4
*Physics	4
French or German.............	4
Civil Government..............	2
Drawing......................	2
	20

THIRD YEAR.

Advanced Algebra............	2
Solid Geometry......	2
English, Literature and Composition	4
*Chemistry...................	4
French or German.....	4
Physiology	2
Drawing. ···	2
	20

*Laboratory practice Saturday mornings.

The above course includes Mathematics, 21¾ per cent.; Science, 21¾ per cent.; English 21¾ per cent.; French or German, 13⅓ per cent.; History, 11¾ per cent.; Drawing, 10 per cent.

DISCUSSION.

Prof. Baker called attention to the fact that only 10 per cent. of the schools tabulated require drawing for admission. He emphasized the value of drawing not only to those intending to pursue an engineering course but also to all others; and urged that engineering schools were greatly in need of advance in this line, and that it would be comparatively easy to secure such advance.

Prof. Tyler inquired whether Prof. Marvin, in the course of his investigations, had, in the first place, procured copies of entrance examinations actually used, or had depended on the statements printed in the various catalogues; in the second place, whether he had considered the general problem of entrance requirements in connection with the courses offered in Manual Training Schools in different parts of the country, expressing the opinion that in the East, at any rate, these schools would become a more and more important factor in the problem, especially as they would in general prepare applicants for the scientific school requirements, but not for those of academic colleges. He referred, also, to the matter of admission by certificate as one that should be considered in this connection. In regard to requirements in Natural Science, he expressed the opinion that such preparation, if well given, was of the utmost value, but otherwise, of very little, if any; and stated that the faculty of the Massachusetts Institute of Technology had always preferred on this account to give instruction in both chemistry and physics from the beginning. Occasionally students who have received fair instruction in chemistry pay insufficient attention to the earlier part of the course for that very reason, and thus, sooner or later, fall below the standard. He suggested that in the case of natural sciences, the scientific schools might to advantage recommend preparation, without requiring it, or holding entrance examinations, since, in general, preparation in natural science, that is, cultivation of the powers of observation, can not be adequately tested by examination.

Prof. Kingsbury remarked that the student of engineering who, at graduation, has a good command of the English language is comparatively rare, while graduating theses bristling with errors in language are the rule. Not only should the entrance requirements be strong, but English should be studied throughout the course.

In the New Hampshire college the faculty has recently adopted the rule that "in marking all exercises English will be taken into account." It is hoped that the rule will lead students into the habits of careful speaking and writing. It is as yet too early to note any results from the step.

GRADUATE AND POST-GRADUATE ENGINEERING DEGREES.

By PALMER C. RICKETTS,

Director of the Rensselaer Polytechnic Institute, Troy, N. Y.

The request of the secretary that the papers be made short is particularly appropriate for the topic now to be discussed. It hardly seems necessary to occupy much time for its consideration.

A degree, conferred at the conclusion of a prescribed course of study, merely signifies that the course has been finished to the satisfaction of the corporation conferring it. Since it means this, and nothing more, it seems appropriate to use that succession of words for the degree, or that succession of letters for its abbreviation which most nearly indicates the character and object of the curriculum.

It is the general custom for engineering schools in this country to confer the degree of Civil Engineer upon candidates who have satisfactorily finished a course, the object of which is to fit them, as far as is practicable, for the practice of the profession of civil engineering; and likewise the degree of Mechanical Engineer at the conclusion of a course in mechanical engineering. Possibly it may be more logical, under

59

the circumstances, to confer the degrees of Bachelor of Civil Engineering and Bachelor of Mechanical Engineering, but the first method has the prestige derived from age and general usage. Objections of some strength would evidently have to be advanced to a board which for sixty years had conferred the degree of Civil Engineer, before any change would be made. Those known to me do not seem to be strong. The argument has been advanced that, since all graduates of such schools do not follow the profession, professional degrees should not be conferred upon any of them until they have had practical experience. I can see no reason in this objection. It matters nothing if some manufacturers, lawyers or ministers have the right to place C. E. or M. E. after their names. It has also often been said that such degrees should not be conferred upon graduates without experience, because the general public in some parts of the country regard the possession of one of these titles as evidence that the young man is a practical engineer, and therefore employ him to take charge of work of great magnitude to the consequent detriment of his more experienced professional brethren and injury to the profession at large. If this belief exist, it is reasonable to suppose that it will not long continue, for such a mental condition would probably cause such a public to place the same confidence in young practitioners of the medical profession, and the doctrine of chances renders rational the supposition that the public would, in consequence, soon be removed.

Against the use of the abbreviations C. E. and M. E. there has been advanced the objection that non-graduate civil and mechanical engineers sometimes use them instead of calling themselves simply civil or mechanical engineers. I can hardly believe that many reputable engineers of any prominence are ignorant of ordinary usage in this respect. The disreputable ones can, of course, make use of any abbreviations they may wish without legal interference, and, in my opinion, without any particular injury to the profession.

As a matter of fact, the whole subject is of minor importance; as long as institutions hardly above the grade of preparatory schools continue to confer engineering degrees, it is of small use to discuss what the character of the degree should be. It is the course, not the degree, which counts, and employers searching for a young graduate ask, "What course did he take in what school?" and, "What kind of a man is he?" not, "Does he have the right to place C. E. or B. C. E. or B. S. after his name?" Later they are only concerned about his experience, and even the school becomes of small importance.

In any case the question is of importance only to the young engineer. It is very unusual, except in the title page of a book, to see the name of an engineer of any prominence followed by an abbreviation indicating that he has taken a degree in a professional school. The abbreviation of the recognized engineering societies displace those of the schools.

I can see no reason for conferring the degree of

Bachelor of Science at the conclusion of a course intended to lead toward the practice of any branch of engineering, no matter how elementary such a course may be.

Master of Engineering is, perhaps, as good as any post-graduate engineering degree. No one would have the right to assume that such a degree meant anything else than that its recipient had taken a prescribed course of study. Its abbreviation would probably likewise be displaced by those of recognized engineering societies.

SECOND PAPER.

By GEORGE F. SWAIN,

Professor of Civil Engineering, Massachusetts Institute of Technology, Boston, Mass.

In discussing the degrees given by engineering schools it is desirable, in the first place, to consider what a degree really means. Probably few teachers will dispute the statement that a college diploma, strictly speaking. is nothing more than a certificate of work done—a statement that the recipient of the degree has satisfactorily performed the collegiate work required for its attainment. This, and this alone, is what is meant by a collegiate degree.*

------

*With purely honorary degrees we need not here concern ourselves. These, however, if properly conferred, represent, similarly, work done, or eminence attained by work done, but outside of the institution conferring the degree. Frequently, however, they simply mean that the college, for some reason or other, chooses to honor the recipient.

A diploma is not a certificate of character. Neither is it a guarantee that the holder is qualified to do responsible work in his profession at once. He may, or may not, be able to do this. Nor is it a recommendation from the institution conferring it, although in some cases the wording upon the parchment states it to be such. Every experienced teacher, however, knows that an institution of learning is occasionably likely or obliged to award its diploma to men whom the members of the Faculty might not personally feel able to recommend. In any class of students the line separating those who receive the diploma and those from whom it is withheld can not, in the nature of things, be sharply drawn; and there is always likely to be a man, or several men, who, while they have technically done the work required, are nevertheless close to the line; and they may be men who will never succeed, and whom the members of the Faculty cannot individually recommend. A man may do the work required for a degree, but may be possessed of personal characteristics which are extremely objectionable; he may be of good ability, but so lazy that he has performed *just* the work required and no more, or he may have taken double the usual time to do it; or he may be stupid and of low mental calibre, but earnest and persevering to an extent offsetting his defects just sufficiently to enable him to receive the diploma.

It is true that in considering a candidate for graduation the personal opinion of his instructors must frequently be the final criterion. Every teacher knows that the results of examinations possess only fictitious accuracy, and that the poor student will frequently obtain a higher mark than the good student. One of the most important recent innovations in methods of teaching, in my opinion, is the tendency to abandon formal examinations, and to make the rank of the student depend entirely upon his term's work, which, of course, must be systematically recorded. In considering a candidate for graduation, therefore, the ultimate criterion will frequently be the answer to the question, "Will he do the institution credit after he leaves it?" or, "Can we recommend him personally?" Nevertheless, these considerations do not affect the validity of the statement that an institution is apt at any time to graduate a man who has technically performed the work required, and from whom the Faculty can not consistently withhold the diploma, although as individuals they might be unwilling to recommend him as an assistant to an engineer applying for one. It is true, of course, and it must not be forgotten, that in an institution where the moral standard and the standard of scholarship are high, where the required work is severe, and where the daily work of the student is carefully followed and deficiencies promptly pointed out—it is true that in such an institution a four years' course will generally result in weeding out the lazy and stupid men, and in most cases those who are personally objectionable, particularly those who are objectionable on account of bad habits. In such a case, therefore, a diploma will, in the mind of the public, come to be a recommendation of the person holding it, because it will be known that only a good man could graduate from such a school. This, however, is simply an inference from the degree and not a characteristic of the degree itself, and I venture to say that even in such a school cases will sometimes occur in which the degree will be awarded to men from whom the Faculty would be very glad to be able to withhold it consistently. If it is true then, as I believe it is, without question, that a collegiate degree simply stands for a certain amount of academic work done by the recipient, and not as a recommendation or as a certificate of character, it would seem desirable that the form of degree conferred should be one which should by its name indicate its true meaning, and show it to be an academic degree, not one likely to be confused with the name of a profession or occupation.

The degrees in common use belong to three classes. On the one hand there are degrees similar to the degrees of A. B., A. M., Ph. D., M. D., which have been awarded for centuries and which are simple abbreviations of the Latin titles: A. B., *Artium Baccalaureus*; A. M., *Artium Magister;* Ph. D., *Philosophiæ Doctor*; M. D., *Medicinæ Doctor.* These are all degrees which have come down to us from the middle ages or from earlier times, and which indicate in themselves what they stand for. For engineering or scientific schools the corresponding and strictly analogous degrees would be S. B. or B. S., *Scientiæ Baccalaureus;* S. D., *Scientiæ Doctor;* or Ph. D., *Philosophiæ*

5

Doctor; the word *Philosophia* being interpreted in its broad general sense. These titles or degrees have no meaning except as academic degrees, and would never be used in any other sense in writing or conversation. A. B. means simply the first or lower, the *baccalaureate* degree in arts, and so with the others. By common usage nowadays the word doctor has come in the popular mind to stand for a Doctor of Medicine, so that the degree M. D. is perhaps, by many, considered simply to represent the name of the profession. Such, however, is not its derivation or its real meaning. We all know the word doctor is much broader in its significance, and includes all sciences or branches of human knowledge —medicine, philosophy, law, music, etc. Doctor means simply a learned man, or one qualified to teach—a teacher.

On the other hand stand the degrees which may be called descriptive degrees, which are of comparatively recent origin, and which have been conferred principally, if not exclusively, in this country. These degrees are simple abbreviations of the name of the profession or occupation for which the recipient is supposed to be prepared: C. E., Civil Engineer; M. E., Mechanical Engineer, or Mining Engineer; D. E., Dynamical Engineer; E. E., Electrical Engineer; S. E., Sanitary Engineer; and very likely there are others of this character. The writer believes that degrees of this kind are in all respects inferior and less suitable as representing academic work done than degrees of the class first referred to. The reasons for this belief are four in number, and may be summarized briefly as follows:

First, the descriptive technical degrees are not in line with and do not correspond to the ordinary academic degrees; and, while not essential, the writer considers it at least desirable that collegiate degrees of all classes should be chosen according to some uniform system. It seems somewhat inconsistent for a university to confer in its classical course (even though the use of options may make it possible to complete such a course without study of the ancient languages) the degree, A. B., and in its technical course the totally dissimilar degree C. E., or E. E., even without reference to the fact, which will be again alluded to, that such technical course does not and can not itself make a man a civil engineer or an electrical engineer. The distinction, however, between the ordinary classical degree of A. B. and the scientific degree of S. B. is plain and obvious. The same is true with regard to the classical post-graduate degrees, A. M. and Ph. D., and the scientific post-graduate degrees M. S. and S. D. The Ph. D. degree may be fairly considered as broader and more advanced than any of the others, and may reasonably be applicable to advanced work in either classical or scientific departments.

The second objection to the descriptive degrees has already been alluded to, namely, the fact that they are simple names or abbreviations of the names of occupations or professions, and their use leads to confusion, especially in schools where the number of courses is large. If a graduate in civil engineering receives the degree C. E., the graduate in electrical engineering the

degree E. E. and so on, what degree shall the graduate
in architecture receive, or the graduate in chemistry?
It would be just as consistent, and just as reasonable,
for the degree in architecture to be the simple letter A.
and the degree in chemistry the simple letter C.—standing for architect and chemist respectively. Just as
reasonable, too, would it be, if this system were carried
to its logical outcome, for doctors of medicine to receive
special degrees according to the particular branch of
medicine which they pursue. Thus, we might have
John Smith, E. D., Eye Doctor; Thomas Jones, L. D.,
Lung Doctor, and so on. Probably no one would for a
moment advocate degrees like these, yet they are no
whit more unreasonable than the multiplication of
purely descriptive engineering or scientific degrees.

The third objection to a descriptive degree is that
it is liable to abuse. Plumbers are nowadays blossom-
ing out in great luxuriance into sanitary engineers.
Many a man practicing as a civil engineer, and entitled
by his attainments to call himself one, finds it desirable
to abbreviate the name of his profession, and to write
C. E. after his name; and many men eminent in their
profession, though without college training, sometimes
find it convenient to do this. Moreover, the temptation is great, to one who is unscrupulous but pushing,
to assume this title as a deliberate means of deception,
under the pretext of using a simple abbreviation. The
temptation is not equally great to appropriate the
degree Bachelor of Science. Legal regulation with
regard to engineering practice, while advocated by

some, is not likely to be enforced for some time at least,
nor do I believe that any movement of this kind would
be strenuously urged by the majority of engineers. In
this respect the engineering profession stands in a position radically different from that occupied by the medical profession. Medical practice laws have in recent
years been enacted in many states, making it illegal for
a practitioner to put the title M. D. after his name
unless the degree has been conferred upon him by a
reputable college; but even this, I am informed, is not
a statute which the regular physicians themselves were
specially desirous of having enacted. It has resulted
from a movement of the public, rather than from a
movement of the medical profession; and certainly the
public necessity for such movement is much greater
than it could ever be in the case of engineering. A
transient visitor in a city, suddenly taken ill and
compelled to call in the services of a physician at a
moment's notice, needs some assurance that the man
he employs has at least had the training which the
degree M. D. would imply. It is a matter of life and
death, and the decision must be made at once. Time
can not be taken to inquire as to the standing of different men. In the case of the engineering profession
there exists no such necessity for immediate action. A
person desiring to employ an engineer has generally
thought the matter over for some time and has had
opportunity to consider the qualifications of various
men, or to bring a competent man from even a distant
place. Moreover, the case is here seldom one of life

and death, and it will rarely occur that a transient is called upon suddenly in the middle of the night to employ an engineer. The abuse of the title or degree C. E., therefore, (and the same holds good of the other similar degrees), is not by any means as serious a matter as the abuse of the degree M. D. Its regulation by law is not a matter of great importance, and is not specially desired by the profession; for every one must realize that an engineer in active practice soon comes to stand upon his known merits and is judged according to the work which he has done and the record which he has made in practice, so that in many cases a degree of any kind soon becomes practically superfluous. Nevertheless, the abuse referred to constitutes a valid, and by no means trivial, objection to a so-called descriptive degree.

The fourth objection to such a degree may be found in the fact that in reality a school training does not make a man an engineer. To confer upon him, therefore, the degree C. E. or M. E. is in itself a contradiction in terms. In order to become an engineer, a man needs much more than the training given in any institution. He needs experience and judgment, as well as knowledge. Experience and judgment can be taught only to a very limited degree in a school. From this point of view, therefore, the simple degree in science, Bachelor of Science, or in philosophy, Bachelor of Philosophy, indicating simply and solely that the man has taken his degree for a course of study, appears very much more suitable than a descriptive degree. To this it may perhaps be replied that the technical school qualifies a man to practice as an engineer to the same extent that the medical school qualifies a man to practice as a physician, and that there is, therefore, as much justification for the degrees C. E., etc., as there is for the degree M. D. The first statement may be true, but it must not be forgotten that the degree M. D., no matter what it has come to mean in the public mind, in reality and by derivation simply means, a man learned in medicine. It is one of the oldest degrees, and indicates by its name and derivation an academic training simply. Similarly, a man who is a bachelor of laws is not necessarily a lawyer.

Intermediate between the simple degrees S. B. or Ph. B. and the purely descriptive degrees C. E., etc., are the degrees such as B. C. E., B. M. E., etc., which are awarded at several schools. A good deal may be said in favor of degrees of this kind, and not much against them. Their name indicates them to be academic degrees; they are not simply the names of professions or occupations; they are in line with the old established classical degrees, and they are not liable to abuse. Their use, however leads to some confusion. and seems to me inadvisable in schools where the number of courses is large. If the graduate in civil engineering receives the degree B. C. E., the graduate in architecture should receive that of Bachelor of Architecture which would have to be abbreviated to B. Arch. in order to distinguish it from the degree B. A. However, the matter of abbreviation is not serious, and can

easily be arranged. There appear no advantages, however, and many disadvantages, in conferring a separate
degree for each course; and the broader degree S. B. or
Ph. B., which is simpler and more general, seems prefer-
able. These degrees indicate, moreover, that the training
in our engineering courses is principally a training in
the scientific principles which underlie the work of the
engineer. Engineering is in itself a practical profession, but is underlaid by a body of scientific principles
without which an engineer stands at a great disadvantage. Nevertheless, the essential practical character of
the profession itself is shown by the fact that many
men have attained eminence as engineers—though perhaps not the highest eminence—with little or no
knowledge of mechanics or mathematics. The training
given in engineering schools is largely or principally a
training in science, and it seems to me an advantage
that the degree awarded should emphasize this fact. I
prefer, therefore, the undergraduate degree of S. B. (or
B. S.); and for advanced degrees, correspondingly, M.
S. or D. S., while if a second and higher advanced
degree is desired, that of Ph. D. appears the broadest
and most suitable. The principal point which I
wish to advocate is that the so-called purely descriptive degrees, like C. E., M. E., when awarded for academic work, are misleading and unsuitable. I can not,
however, feel that the matter of engineering degrees is
one of such supreme importance as it is by some considered, for the reason already stated, namely, that a
man in practice very soon comes to stand upon his own

acquired reputation, and is judged with reference to his character, his experience, and the record he has made in practice, frequently without any reference whatever to his academic degree, if he has one. At the same time, schools must give diplomas as certificates of work done, and my own feeling is very strong that there is no consistent, rational, or reasonable ground for conferring a so-called descriptive degree.

Before closing, mention should perhaps be made of the practice which is in vogue in some schools, by which the undergraduate degree is Ph. B., or S. B.; the advanced degree of C. E. being given after the recipient has had a certain amount of experience in actual engineering work. This practice is certainly much more justifiable than that of awarding the degree C. E. for undergraduate work alone; it may obviate the fourth objection made in this paper, namely, that school training does not make an engineer. It does not, however, obviate the other objections, and the degree of C. E. thus conferred is practically an honorary degree, in the sense that it is not conferred for academic work, but for work done *in absentia*. While the discussion of honorary degrees may not properly be included in the title of this paper, my views regarding them are radical, for I believe we should be much better off without them, unless there were some mark by which to distinguish an honorary degree from a degree representing actual college work. It seems to me there is great danger of cheapening degrees by the too liberal and promiscuous giving of post-graduate degrees without work done at the institution. If such degrees could be limited to those who, by character and attainments, fully deserve them, and if they could in some way be distinguished from academic degrees, their use might lead only to a healthy stimulus toward research and study; but there is no doubt that in some cases they have been awarded on far different grounds; for instance, as a recognition of mere wealth, influence, or perhaps political position, independent of scholarly acquirements or work done anywhere.

If the object of a school is to give its degree to as large a number of persons as possible, or to solicit support and interest by thus spreading its honors, it will naturally be liberal in granting post-graduate degrees without actual instruction or academic work done. I believe, however, that the practice is easily carried to excess. I believe, moreover, that the object of a school is simply and solely to give instruction, and to award diplomas for proficiency attained under that instruction, as well as to aid in the advancement of learning by means of original research; not to have its faculty act as a board to pass judgment on the eminence or attainments of men whom they may never have seen. I may state, as an example of great conservatism, of which 1 cordially approve, that the institution with which I have the honor of being connected has never, so far as I am aware, awarded a diploma for work done *in absentia*, or for anything except as a certificate of actual work done at the school or under its direction. In no other way, it seems to me, can the degree of a school be made to represent something definite and reliable.

GRADUATE AND POST-GRADUATE DEGREES.

By ROBERT H. THURSTON,

Director of Sibley College, Cornell University, Ithaca, N. Y.

The designation of the degree to be awarded the undergraduate at the completion of his course in any engineering school has been a prolific source of discussions and even of dispute for many years. The title to be assigned the graduate student completing an advanced course in these schools has only been a less widely discussed matter because such courses have been of later origin and much less generally offered by professional schools of engineering. From the first, many schools have followed the course of the older non-professional, the purely educational colleges, and graduated Bachelors of Science in special lines of work; others have simply labeled their graduates "Civil Engineer", "Mechanical Engineer"; still others have adopted the hybrid title "Bachelor of Engineering". The second degree is sometimes "Mechanical or Civil Engineer", sometimes "Master of Science" in one or the other branch, sometimes "Master of Civil" or of "Mechanical Engineering". The doctorate, so far as the writer is aware, has never been offered in engineering except by a single institution, and as an honorary degree, and then with exceedingly great caution and very rarely. In a few instances the title conferred is entirely different from either of the old forms, as "Dynamic Engineer"; to which designation the complimentary title "Static Engineer" has seldom, if ever, been added. Choice has apparently been usually determined by force of example, as where the custom of the older schools is followed, by professional *esprit*, as where the title given is that of the profession itself, or by the spirit of innovation, as where the title is newly invented for the occasion. Occasionally, as in the case of Stanford University, the custom is established for all schools and courses alike, by the general faculty; and all graduates are dubbed A. B., whether in arts, sciences, literature, or in engineering, thus giving perfect democracy among alumni, and by the same act, taking from the degree all value for the professional, except as indicating his graduation from a reputable college. In this case, the initials of the college would perhaps constitute a still better badge.

The writer was compelled to take up this question and to promptly decide for himself as long ago as 1871, when called upon to take leading part in the establishment of a course of instruction in mechanical engineering, intended to be as distinctive in its field as was, and is, that offered at the Rensselaer Polytechnic in civil engineering. A thoughtful and careful discussion of the subject with the then best-known and most competent members of the profession confirmed his own impressions and led to the selection of the professional title rather than that of the older class of schools and colleges. The latter, having at least the merits of novelty and logical correctness, was rejected simply as not likely to find favor with either the followers of the gymnastic schools or members of the profession.

The reasons for the final decision are simple and easily summarized:—The school to be established was intended to be a professional school, distinctively. It was important that the degree offered should, if practicable, indicate that fact and give some presumption that the student graduating from its professional course might be expected to exhibit some special fitness and a preparation for entering and advancing in that profession. It was desirable that neither the school, the course, nor the graduate should be confounded with those schools, courses, and graduates so nearly universally recognized as below the standard set by the profession—the schools organized in connection with the older institutions of learning, controlled by non-professionals, by the clergy largely, and offering singularly inadequate courses of instruction; graduating students neither educated, nor professionally trained, hybrids comparatively weak in educational branches and usually compelled to unlearn much of their "professional" instruction before they could be entrusted with any really useful office or field work.

At that time, and in nearly all such schools, the attempt was being made under the pressure of the non-professionals in control, whatever the views of the engineers nominally in charge, to give at one and the same time, a college education and a professional training in four years, sometimes in three; notwithstanding the now recognized and obvious fact that either education, to be satisfactory, should occupy the full term and demand steady and earnest application throughout.

The fact that the student must either get his education first and his professional training later, as in law, medicine, or any other profession, or must choose either the one or the other, was not then as generally admitted as now, and the most singular mixtures of literature, history and other non-professional studies with engineering were often prescribed; as where, in one now famous institution of learning, "biblical exegesis" constituted a portion of the regular course in engineering, or where, as in the early days of Cornell University, Roman history was similarly imbedded in a course nominally that in civil engineering, "like a fly-speck on a white wall" as the finally emancipated head of the department was accustomed to say. In the same institution, in the earlier days, we had a course of instruction in "English" for years; taking the place of professional work in mechanical engineering, injuring the efficiency of the course while giving practically little advantage as literary training. This subject is now obtained with better results in the preparatory schools, and the consequent elevation and improvement of the course, now demanding more of preparation before entering gives an average student better literary standing at entrance than he formerly had at graduation, and at the same time permits his securing a comparatively satisfactory, and truly professional, training. It was recognized, finally, that a professional training is not an education, in the correct and accepted signification of the term, and that the best obtainable education should precede the work of the professional school.

Each should do its work independently if it is to be done well. Either, attempting the work of the other, must prove more or less of a failure in proportion to the fraction of time given to the foreign element. A good professional school, devoting all its time to its legitimate work, still finds that it has no time to spare, and usually that more time, still, would be acceptable. For these hybrid courses of the older regime the older designation was recognized as appropriate enough. They were not, properly speaking, professional schools; they were, in considerable degree, schools of applied sciences. With them Bachelor of Science was as appropriate as a title as for the schools of pure science beside them.

It was thought by the strongest men in the profession that the professional title would prove more acceptable for the distinctively professional school, both as being more appropriate in view of the more nearly professional nature of the school, its closer approximation to the standard set by the other professional schools, as of law and of medicine, and as being more likely to satisfy the demands of the student and alumnus, a matter in itself of some importance. It was also thought by many that the old Latin term, bachelor, was hardly consonant with modern and popular ideas; the classical and somewhat incongruous shade of tone being likely to strike unpleasantly upon the ear of anyone at all inclined to be critical in matters involving literary taste and accuracy. "Bachelor of Science in Engineering" was not so bad; but "Bachelor of Engineering" seemed to many entirely inadmissible.* For the second degree, however, the good old English term, "Master," awakened no opposition, and a "Doctorate in Engineering" was admitted on the ground that Engineering was coming to be recognized as a learned profession and was actually demanding more of its practitioners in its scientific preparatory work than the other professional schools and its highest order of practice might well be considered to entitle the practitioner, thus standing at the head of his profession or in its front rank, to the designation of Doctor.† The popular assumption that the title is confined, properly, to the profession of medicine has no basis in derivation or in practice.

It was these considerations, mainly, which led the writer, personally, and, he thinks, the majority of the abler members of the profession and the acknowledged leaders of the time, to agree upon the use of the title of the profession for a first degree and to adopt a Master's degree for a second and the Doctorate for the highest degree proposed.

---

* The term bachelor is from the Latin, baccalaureus, one crowned with laurel. In the French it becomes "a young squire, not made a knight " Its first English meaning was "a young, unmarried man." In old times, the student undergraduate was forbidden by the law of the universities to marry, on pain of expulsion. Violation of this law by Wm. Lee resulted in his invention of the stocking loom.

Master is from the old Anglo-Saxon, maester, one who has attained physical superiority over other men; later, one who is superior in any art, profession, science or department of learning. The first of these two collegiate titles is, in the opinion of many members of the profession, as incongruous for the engineering schools as the second is appropriate.

---

† Doctor: From the Latin, doctor, doceo, doctus, to teach; designating, in English, one who has received the highest degree from an institution of learning; one who is learned, an expert, and an adept; a teacher of his craft.

On the establishment of the course offered in 1871 at the Stevens Institute of Technology, so far as professional, by the writer, then the professor of engineering in the newly organized school, the undergraduate course led to the degree "Mechanical Engineer"; while the advanced courses were left to be established later. The honorary degree of Doctor was in a few instances conferred. Postgraduate courses had not been established at the time of the transfer of the writer to Sibley College, Cornell University, in 1885. When taking the directorship of this institution, the writer was authorized and directed to organize and establish its courses of instruction, to create departments of study and professional work and to select and nominate the incumbents of the several chairs; in fact, to completely organize a school of mechanical engineering, and to set it in operation. The same considerations which had determined the partial adoption of the scheme of 1871 at Hoboken induced the recommendation of a similar scheme for the new college. The titles proposed for the first and second degrees were adopted. The doctorate has not yet been established, although large numbers of graduate students are working for the second degree and the indications seem to be favorable to the experiment of establishing the higher course of professional work, possibly a three years' course *in absentia*, with commutation of one year if worked out entirely in the college under the immediate supervision of its Faculty.

In 1893, ninety-three took the first and fourteen the second degree; in 1894, the number taking the

6

second degree was seventeen; the total being a little
less than in 1893.   In 1893 and 1894, over sixty candidates were on the lists for second degrees in 1893 and
later.   Many were instructors taking three or four
years to perform the work, being seriously impeded by
their daily duties.   The number graduating in 1886,
the year in which these courses and degrees were first
put in effect, was five, taking the first degree; in 1887
sixteen took the first and three the second degree, and
from that time on the growth of the institution was
exceedingly rapid, and was attributed, in part, to the
wisdom shown by the Faculty in the adoption of the
professional title for its degree.   Of this, of course,
there can be no really crucial proof; but it was probably one of numerous conspiring causes.

There are various objections urged against this
system of nomenclature of degrees; some of which
undoubtedly have weight, some of which are as unquestionably farcical.   That most commonly and
most seriously urged, perhaps, is the undesirability
of conferring a degree which is at the same time
the designation of the profession itself.   This seems
to the writer rather an argument for than against
the title chosen.   The undeniable fact that the
graduate is only prepared to begin to learn the
essential practical routine of his vocation, and is not
and can not be prepared to practice, has no more weight
than in medicine, where every graduate is a "doctor."
The fact is well understood by everyone that his title,
as conferred by the school, is simply an assurance that
he has had a course of professional instruction, and is

thus given a certain indispensable preparation for entrance into the profession which he has chosen, precisely as in any other profession. The degree "Civil" or "Mechanical Engineer" gives no more presumptive evidence that he is competent to practice than does the degree of "Doctor of Medicine" in that field. Neither ever is or ever will be misunderstood. It is just as true that a master's diploma in science, literature, arts or engineering gives no assurance that the holder is a master in the vocation he may have selected; it is simply the certificate of a reasonable proficiency in those branches of learning which are customarily pursued in such courses as are prescribed as leading to the stated degree. The same is as true of the doctorate in any branch or profession. No one ever mistakes these diplomas for certificates of proficiency in anything outside the courses of the schools to which they each specifically appertain.

It is the business of the schools of the professions to make certain that these diplomas, however, represent as strong, condensed, and fruitful a course, each, in the sciences underlying the profession, as the state of contemporary science and professional learning and practice permits—that is to say, so much of human knowledge as bears upon that vocation in the form of the history of the development of the art and its state at the time; the applied sciences so far as they bear upon professional work; the literatures of our own and other nations so far as they have professional importance; the methods of allied arts so far as they can properly be described and illustrated in the lecture room, class-room and laboratories, and the theory and practice of scientific research so far as bearing upon the problems arising in practice or in the development of the sciences finding application therein. In many cases, even the practice of the profession in certain important lines may be taught and illustrated; and to that extent the graduate is often better prepared for business than his older and less favored colleague, who has never had the advantages of systematic instruction and laboratory practice.   It is the business of the professional school to develop methods of reducing the work of the practitioner to scientific form and method, and to that extent to teach the practice as well as the theory of the art.

It is in this manner that the methods of scientific determination of the efficiency of steam-engines, boilers, and other apparatus of the engineer has come to constitute a part of every course of instruction in any truly professional school.   The chemistry and physics of the development, and transfer and storage and transformation of heat in the production of mechanical energy is thus supplemented by the engineer's practice in measuring the useful effect obtained from a stated quantity of thermal energy thus derived and dealt with.   In engineering, the schools are schools of applied science, and it is their purpose and duty to make the instruction in application as extensive and complete as the state of the sciences and arts permits, quite as much as to give a knowledge of the underlying pure sciences.

To dub the graduate of a professional engineering school Bachelor of Science, or those taking advanced courses Masters of Science and Doctors of Science, seems as inaccurate and unsatisfying as would be the adoption of the same system in many other professional schools. Law and medicine are based upon sciences and their practice is a system of applied science; but the distinction between the student of pure science and the professional is wisely observed by emphasizing the professional side, that of application; and the doctor in medicine or in law, just as much a scientific man as his neighbor, the engineer, is designated by terms which leave no possibility of confounding him with the chemist, the physicist, the physiologist, the biologist, whose learning he must always borrow for his professional work. Similarly it would seem that the engineer should be distinctively designated as an expert in scientific professional work, not as a man of science simply. John Doe, M. E., or Richard Roe, C. E., is unmistakably marked professionally; John Doe, B. S., or Richard Roe, B. S., presumably a student of sciences, is certainly not likely to be taken by the stranger reading his card as legitimately inducted into the profession which he may claim as his.

Perhaps the most potent argument in favor of the adoption and retention of the special title is the fact that a very large proportion of the graduates of engineering schools, and an increasing proportion, are carrying that title. Another important consideration is the fact that the recipients of the degrees given prefer the professional title. When, with the advance in the

requirements for entrance and the considerable accompanying improvement of the professional courses at
Cornell, some years ago, the title was changed from
Bachelor to Civil and Mechanical Engineer, it was provided that for the time either title might be received,
at the option of the graduate, in electrical engineering
courses, not one graduate, out of scores taking that
course of study, ever called for the degree of Bachelor
of Science. All preferred the degree giving professional
distinction, precisely as in other professional schools.
That provision still stands, but it has completely dropped
out of sight through non-application. The young graduate aspires to be known as a member of a profession
and an aspirant in engineering. not as a student in science, simply, however honorable and honored the latter
vocation may be. His pride lies in professional success,
and all his hopes, ambitions, and labors tend that way.
Even the title assigned him by his *alma mater*, intrinsi-
cally unessential as it in fact is, becomes to him a matter of interest and pride, and assumes real importance.

This form of diploma is preferred by the greater
number of the representative men in the profession.
They welcome the young engineer into the profession,
and adopt him into the society, not as distinguished as
a student in the sciences, but as one whose ambitions lie
in the same line with their own, as one who aspires to
follow in their footsteps, to emulate and improve upon
their work, to accomplish all that real talent, genius,
education, industry will permit in what their seniors regard as the noblest of the professions, the most useful

and fruitful of direct good of all the vocations. They welcome him as a novice in engineering, and take him into the profession as one of their own family. The school is simply the first stage of professional work, and its title should indicate that fact.

The question of designating the degree conferred is, after all, a small matter beside the problem which is involved in the construction of a suitable professional course of instruction for the real professional school. By real professional school is here meant an engineering school in which the work is purely that of professional instruction and preparation, precisely as in any real professional school of law or of medicine, and in which no working time is sacrificed to general education, to "culture" or to purely gymnastic studies. Its requirements for admission are, properly, simply those branches of learning which necessarily preface the work of the professional, as mathematics up to the point at which either the schools from which the candidates for admission mainly come cease to teach the higher mathematics, or the work of applied mathematics of the science of the profession properly begins. These requirements do not properly include any branches not finding later application either directly in professional work or as introductory to studies or laboratory work, forming a part of the professional course. The course itself properly consists of just so much of the sciences, the arts, the literatures of contemporary and earlier times finding application in the practice of the profession, as essential elements of professional work, and so

much of methods of application, as can be systematically given in a course of the length assumed as practicable. In engineering schools, four years is generally
thought none too long for even the purely professional
course; in schools of law and medicine, two years and
often less may be admitted. The engineer has come
to be the most completely trained, the most learned,
among professionals. Given, as is now not uncommon, a good preliminary course of culture, of general
gymnastic education, supplemented by a full course of
professional training, in a real professional school, and
in the higher school of practice, he is necessarily the
most thoroughly educated and at the same time the
most learned of professional men. James Watt was perhaps the leading member of the Lunar Club, composed
of the great scientific men of his time. The modern engineer who has enjoyed all the opportunities coming to
the man of moderate circumstances of our time, and who
has taken full advantage of them, or who, as a "self-
made man," has acquired both an educational and professional training, may always emulate Watt in this
direction. But whatever his location, position or specialty, the ideal and representative member of the engineering profession, hereafter, will be a man of ability,
strength, and supreme integrity, who has secured the
best education that the best university can offer or that
can be obtained by study and travel, perhaps, followed
by the best professional training that the best professional schools can give, and who has shown by his works
that he is a fit disciple of Telford or of Watt. It is of

comparatively little consequence what title shall be conferred by the schools upon this representative engineer.

Reviewing the field, it would seem probable that a variety, both of courses and of titles, must be accepted and endured for a time. Colleges and professional schools alike must usually be restricted and controlled in their work by the possibilities. All seek to make their requirements for admission as high as practicable; all are compelled to accept what they can, for the moment, secure from the preparatory schools. The so-called schools of engineering will probably for years to come, in some sections of the country and under ordinary local conditions of environment, be compelled to offer semi, or partial, professional courses, incorporating with the elementary work of the purely and truly professional, more or less of the gymnastic and educational work of the nonprofessional schools. A few, and perhaps usually the independent engineering colleges, will be able to offer courses demanding the higher mathematics and the modern languages, in part, for entrance, and consisting mainly of professional work and studies in applied science. Now and then one, the numbers probably increasing with the progress of time, may be able to secure full preparation for a purely professional curriculum, and may thus attain the standing of a real and unadulterated professional school. Such technical colleges, whether independent or connected with the universities, must probably long remain few in number, possibly small in magnitude.

The first of these classes of school, with its mixed course, its limited professional, largely educational, curriculum, ought not, in fairness, to receive the title of professional school; it gives simply a course of study which properly takes its place as a modification of the usual and standard higher courses in science, of the nonprofessional colleges and the universities, and its degree should, obviously and naturally, following convention, be that of Bachelor of Science, and the reading of its diploma may be qualified by a statement of the special branch which constitutes its characteristic feature. It would be neither logically correct, nor fair to its graduates or to the profession, to unqualifiedly call this a professional school, or to give its graduates what might be interpreted to be a title to entrance into the profession as from a truly professional school. It would be as wrong, and exhibit as serious incongruity, as to dub M. D. the graduate of a high school in which anatomy, physiology and hygiene had been taught with exceptional development at the expense of the usual and regular high school studies, or M. E. when manual training had been similarly added to the older courses, with an attempt at teaching applied mechanics without adequate preparation in the calculus and accessory mathematics, and without laboratory or other higher training in the mechanics and the physical sciences.

The second of these classes, once it has succeeded in fully emancipating itself from the thraldrom of the preparatory schools, in formulating its courses on a correct basis of applied science, and in making them completely professional, will probably prefer to give a title to indicate that fact; as have for many years already some of the leading schools, even before reaching that higher stage. With these schools, on a level in position and standing with the law schools, and having, as a rule, higher requirements for entrance and a stronger, as well as much longer, course of work in exclusively professional lines, the wise and the politic plan would seem to be to give titles denoting and defining their character with honesty and directness, distinguishing themselves from the schools of mixed curricula as completely as from those of an absolutely nonprofessional kind. The title should be apposite to the work.

In distinguishing between schools of these several grades, it would, perhaps, be well to make some such classification as the following: Where the curriculum includes less than one-half modern languages and applied sciences, and, commencing in the line of mathematics, with elementary algebra and geometry, and terminating with elementary applied mechanics in the senior year requires, for entrance, the common school branches only, and contains no laboratory instruction except in chemistry, I would consider the course as in no sense a really professional one, and would give the degree of Bachelor of Science simply. Where plain geometry and elementary algebra through quadratics are required for entrance on a four years' course, in which one-half or more of the work is in modern languages and the applied physical sciences and in laboratories, and where the applied mechanics—a strong

course in that subject—comes in the junior year, as introductory to professional work in the senior year, I
should consider that we have reached the border line
and would give either the bachelor's degree in engineering, or the professional title, accordingly as the character of the course, in detail, approximates the one or
the other, pure or applied science, most closely. Where
a four years' strong course of applied science and mainly
professional work is offered, its applied mechanics in
the junior or the sophomore year, the higher mathematics being required for entrance, all purely educational and gymnastic study being supplanted by work
directly bearing upon the main purpose of the course,
and with extensive lines of laboratory work, in the sciences and engineering, in the junior and senior years,
the school of engineering becomes fully the equal in
rank with the schools of medicine of the highest class
and superior to the average, and stands above all the
law schools in length and strength of professional
courses, and should unquestionably offer the professional titles.

The practice of the engineering schools seems to be
approximating this classification already, and the
schools giving the first and second forms of curricula
are, in many cases, offering the title given by those approximating the last form, after a specified amount of
graduate work has been performed; while the higher
class of professional school is taking the graduates of
the others for post-graduate work in professional
branches and offering them the appropriate degree, often

supplemented later by its own advanced degree. It is, in fact, not a bad plan for the student desiring to secure a good scientific training in engineering to take his first, B. S., degree in the nearest and most convenient school or college, advancing, after graduation, into the semi-professional school of the second grade, and finally completing his work in the purely professional school, and perhaps even then taking an additional year for laboratory work and research in lines in which he proposes to specialize, taking a master's degree, or its equivalent, in conclusion of his final work. Such cases are not unknown in my own experience and seem likely to become somewhat common hereafter, as the number of students seeking graduate work in engineering is rapidly increasing. (The numbers registered for work of this kind in Sibley College in 1892-3 and '93-4, respectively, out of 560 and 620 students, were 64 and 68. Of these, 14 and 17 took the master's degree.)

Such a course is thought by many, in the profession as well as outside it, to constitute the ideal preparation for work in life; the young man fortunate enough to be able to give the time, and to pay its cost, securing first an education and then a professional training such as will, in the end, permit his easy acquirement, if he have the talent—without which none should enter upon such professional work—of reputation and competence, and enable him to make profitable use of all those opportunities, professional, or, and especially, in culture, which come only to the educated, as well as professionally accomplished, man. This is coming to be a

common plan of education with able and thoughtful
young men, and the number adopting it is rapidly
growing. Students completing the courses in arts and
in letters in our universities are sometimes, and in constantly increasing numbers doing the same thing, and
thus securing, first, an education; second, a professional,
scientific training in engineering. Such men reach the
level of experimental investigation comparatively easily
and quickly, and enjoy, as does none other, that highest
pleasure of combining study with original research in
previously unexplored fields of science and professional
work. Although so long and so powerful a diversion
of the mind from practical matters is apt to give a permanent set to the mind of the student having insufficient talent for work, converting him into the imprac-
ticable theorist, those who have the genius for engineering can not be seriously affected in this manner,
and the right man, in his right place, ultimately profits
enormously by such a training. Opportunities come
late, usually; and he has all the years from his leaving college to the age of thirty-five or forty to fit himself into his place in professional practice.

TEACHERS AND TEXT-BOOKS IN MATHEMATICS IN TECHNICAL SCHOOLS.

By MANSFIELD MERRIMAN,

Professor of Civil Engineering in Lehigh University, Bethlehem, Pa.

As an introduction to the discussion of this subject the following propositions or theses are advanced:

First. In pure mathematics the text-book is more important than the teacher.

Second. Teachers of mathematics in technical schools should be in sympathy with, or have had experience in, practical engineering work.

Third. A revision of mathematical text-books from the technical point of view would be of great benefit to teachers and students, and lead to better results in all directions.

Regarding the first proposition it is sufficient to say that an exercise in mathematics involves careful preparation on the part of the student. The text-book usually contains all that is necessary for students of average ability to thoroughly understand the lesson assigned, and the function of the teacher in the class-room is mainly to keep order, incite the laggards to activity, and see that the students have understood the principles and solved the problems. While a good teacher will do much by his personality to increase the interest of the class, it is, after all, the text-book which molds the habits of thought and methods of investigation. The mathematical text-book is also one of the tools which the student consults and uses during his subsequent course of study and practice, and thus its influence extends over many years. Languages and practical engineering subjects can be taught without books, but in mathematics they are indispensable both for teacher and student, unless the class be very small. It is, therefore, important that the character of text-books should be such as to promote habits of clear thought and accurate computation, and lead to a practical mastery of the methods of mathematical analysis.

The second proposition will seem self-evident to all teachers of engineering. It may be doubted by those who claim that mathematics should be studied for the sake of discipline, or for the sake of the theory itself as a part of scientific knowledge, but this view is not now held by our best educators. Mathematics furnishes no better discipline than can be obtained from the study of Greek, botany, history, or any other subject. Mental discipline comes from hard work, continued thought, and a desire to thoroughly master the subject and discover new truths. In engineering courses of study mathematics is an instrument or tool which the student is to use in actual investigations, and if it be taught from this point of view better results both in respect to mental discipline and actual knowledge are likely to be secured than if practical applications are left altogether out of sight, as is often apt to be the case. It is not uncommon in many technical courses, particularly those given in universities, that teachers in mathematics are employed who are recent classical graduates without practical experience of any kind. One case which has come to my knowledge is that of such a teacher who was given charge of a class of engineering students in analytical mechanics, although he had never studied the subject; it was said that he was quite successful, but how much better might he have done if he had been over a course in applied mechanics and had had experience in its practical applications to structural work or machinery. Nothing lends greater interest to the truths of mathematics than that they can be applied to the promotion of the welfare of mankind, and as engineering is one of the professions having this as its object no one can seriously doubt that teachers of mathematics should be in sympathy with the main lines of technical progress.

The third propoposition naturally follows as a corollary from the first and second, unless it be maintained that mathematical text-books are now perfectly adapted to the needs of engineering students. After consultation with many teachers of mathematics and engineering I have found none who does not think that the books can be improved. A successful revision that shall break away from some of the old established precedents and introduce methods in harmony with lines of actual technical applications would exert a great influence upon mathematical teaching. Such a proposed series would perhaps involve a slightly different arrangement

7

of the various branches and a partial redistribution of
the hours of the course of study. A new series of mathematical text-books when so many are now in the
market would not be seriously suggested unless it was
thought that very great improvements could be made
in the direction indicated. Thoroughly believing this
I offer the following thoughts as to the main features
that perhaps should characterize such a revision.

A series of mathematical text-books for the courses
of study in technical schools, covering all and perhaps
more ground than would be necessary, might embrace
seven volumes: (1) Algebra, (2) Geometry, (3) Trigonometry, (4) Mensuration and Computation, (5)
Analytical Geometry, (6) Calculus, and (7) Higher
Mathematics. The subjects of descriptive geometry and
mechanics will not be considered here, as they do not
properly form a part of a course in pure mathematics. The
subjects of Algebra and Geometry, although generally
required for admission to engineering courses, are included in the plan, since the preparatory schools need to
be especially influenced in sound methods of instruction.

The book on Algebra might profitably differ from
those in general use by omitting or abridging the topics
of ratio and proportion, permutations and combinations, harmonic progression, indeterminate equations,
continued fractions, inequalities, convergence of series,
and everything relating to logarithms. On the other
hand, equations of the third and fourth degrees, the
binomial formula, indeterminate coefficients and im-

aginary quantities should be more fully elaborated, and
be illustrated by many examples and problems. Horner's
method and Sturm's theorem need not be treated in
this volume as the former comes properly in the course
of computation, and the latter can not be thoroughly
treated without a knowledge of the calculus. The graphical representation of algebraic expressions and equations, now quite unnoticed in most books, should be
introduced early and form a prominent part of all
discussions.

In regard to Geometry the text-books in common
use seem to need more revision than those on any other
subject. They are largely algebraic, filled with signs
and symbols, not only for the common operations, but
for such words as parallel, angle, and square, all of
which is not conducive to clear and accurate reasoning.
The subject of geometrical teaching has received much
attention in England, and I think it is the general conclusion that Euclid furnishes the best basis for a
beginner. Personally I have always been thankful that
in youth circumstances threw in my way a copy of
Euclid, and I know of no mathematical book which is
likely to exercise a greater influence in promoting clear
thought and logical reasoning. The proposed volume
on geometry might include the first three or four books
of Euclid, followed by propositions and problems to be
solved by the student himself, while the remaining
books might be arranged according to the more modern
order, preserving, however, Euclid's style of demon-
stration. Geometry is not merely the deduction of

beautiful and valuable truths, but it is a powerful instrument for investigation; most American text-books seem to dwell too much on the former, so that often engineering students are very deficient in powers of geometrical reasoning. A partial return to the method of Euclid might hence be advantageous.

In Trigonometry the work should be limited to plane triangles and the discussion of circular functions, spherical triangles being postponed until later in the course where they can be studied by civil engineers and be omitted altogether by mechanical and electrical students. Tables of natural functions should be the only ones here introduced, since the practice of studying logarithms and Trigonometry simultaneously is objectionable for many reasons. A four place table of natural sines and tangents will well illustrate all classroom problems in Trigonometry and is sufficiently precise for a large part of actual engineering work.

The course in Mensuration and Computation is suggested on account of the marked incapacity of students and graduates to perform numerical operations with intelligence and precision. In such a book the common rules for mensuration would occupy only a small space, but the subject of observations and their errors would receive notice and methods of conducting numerical computations be fully inculcated, so that the precision of a computed result may properly correspond with the data. The theory and use of logarithms would here also find its place, with applications not only to Trigonometry but to numerous other problems. The theory of interpolation in numerical tables, and methods of computing tables would be discussed. Graphical computations and representations might be introduced with advantage, and if thought best, the slide rule might be used for some of the numerical work.

Analytical or Co-ordinate Geometry comes next in order, and here too much space should not be given to the investigation of the properties of the conic sections, but curves of other classes should be freely used. The actual construction of many of these curves from their equations should be made on cross-section paper, the constants being varied so as to show the different species. Graphical methods for the solution of equations would also be interesting and valuable. A book on this subject, however, need not be lengthy, since the full discussion of the properties of curves can better be made by the help of calculus.

In regard to Calculus, it seems to me that differentiation and integration should be taught simultaneously. Integration is not a direct process, and the many bulky volumes on the subject contain nothing more than the algebraic reduction of given differentials to forms whose integrals are known. For this part of the work a dictionary of integrals might be given in which the student could find the integral of a differential by referring the latter to its proper class, genus, and species, perhaps according to a classification analogous to that used in Botany. Engineers and mathematicians always have their books at hand when an expression is to be integrated, and in this case students should have the same privilege. Numerous practical exercises and problems should be given in order that the method of the analysis may be fully grasped, and it is more important that the fundamental methods should be thoroughly understood than that an extended course in differential equations should be rapidly covered.

The last volume of the series, called Higher Mathematics for want of a better name, would include chapters on miscellaneous subjects, only a few of which need be read by any one class. Modern synthetic geometry, spherical trigonometry, hyperbolic trigonometry, determinants, probability, higher differential equations, vector analysis, quaternions, the theory of functions, and perhaps non-euclidean geometry, might be discussed, together with some account of the history and literature of mathematics. A few chapters of this book would furnish an excellent review of parts of the previous course, besides giving an introduction to methods which are daily coming more and more into use in technical investigations.

In conclusion it may be said that the above thoughts are presented with the hope that the time may come when students in technical schools will pursue their mathematical studies with greater interest and better results than at present. Having never taught pure mathematics I can not represent at all the opinions of professors of that science, but from consultation with many teachers of engineering I feel that some views here expressed will meet their approval. It will not be possible for me to attempt the preparation of a series of books on the plan sugggested, but if some of the younger members of this Society would take the matter in hand much good work might be done for the promotion of sound mathematical education in technical schools.

ANOTHER VIEW OF THE CALCULUS.

DISCUSSION—(by letter.)

By WM. W. CARSON.

Professor of Civil Engineering, University of Tennessee, Knoxville, Tenn.

One of the most able, as well as cultured, of American engineers has said that he has no choice as to theories so long as the calculus gives true results. But a teacher should take other ground. He should seek to make every tool he puts in the hands of his students as efficient as possible by showing. if he can, just what it is. But he should refrain from warping their judgments and frittering away their time and energies by seeking to prove to them that which is not so. I hold that false demonstrations, and lack of idea of what the calculus is, cause our students serious loss. It is a singular fact that, while writers tell us what Algebra, Geometry, Trigonometry, etc., are, they do not (as far as I am aware) tell us what calculus is. I think

that for two hundred years we have used this branch
of mathematics without knowing its real nature. I
think the failure to recognize the fact that it deals
mainly with a new class of quantities has produced all
the trouble. Evidently a resort to "ways that are
dark" is a necessity in proving of one class of quantities what is only true of another. I think that the
new quantities are hypotheticals—that is, that they are
not quantities that do exist, but quantities that would
exist under certain conditions. We certainly use such
quantities, though almost unconsciously, as the effort
to give precise definitions of such quantities as
velocity, density, intensity, etc., will show. Thus,
the density of a body, at a given point, is the quantity
of matter a unit of volume would contain if the conditions pertaining to that point were maintained. In
my judgment mathematics never discusses anything
but finite quantities. I think that, in mathematics,
the finite and the infinite is a classification both
needless and obstructing. I think that the true classification is the real and the hypothetical. I think that
the calculus is the branch of mathematics that connects
these two classes of quantities. Differentiation passes
from the real to the hypothetical. Integration passes
from the hypothetical to the real. Only one hypothe-
sis is ever made in the calculus—that of maintenance
of conditions as will be seen when I come to define a
differential—and so its hypothetical quantities are
simpler than the real, and easier to find. The hypothetical being found, the finding of the real is simply a

matter of integration.   This explains the great power
of the so-called infinitesimal calculus, and suggests
what the user of it has always really done.   A careful
scrutiny of his work will show that, in getting a differential, from the conditions of the problem, he always
takes a typical set of the conditions obtaining in the
magnitude ultimately sought and then determines the
magnitude that would exist if these conditions did not
change.   This is what he really does, and it is all that
is essential.   But the hypothetical magnitude so found
differs from the real magnitude which he thinks he
ought to find, and so the supposed error is covered up
by the introduction of the infinitesimal idea.   The sole
office of this last is to throw a haze over the whole
matter by making the magnitude too intangible for the
mind to take good hold of.   You will readily see that,
with the idea of a differential which I am about to
propose, the equations cease to be approximations and
become to the last degree exact.

   To find the connection between any magnitude and
its differential: Let x stand for any variable quantity
whatever, either dependent or independent, and let
$F(x)$ stand for any function of x whatever.   Let the
conditions to be maintained be those which obtain in
$F(x)$ for some typical value, a, of x.   I define the
differential of $F(x)$, corresponding to this value of x,
to be the change that would be wrought in $F(x)$ by
any change dx in x under the conditions obtaining in
$F(x)$ for the value a of x.   To find this differential:
Let x pass from a to b.   The change $b-a$ in x produces

the change $F(b)$—$F(a)$ in $F(x)$ under all the conditions that obtain between a and b. If we divide this real change in $F(x)$ by the number of units in b—a we evidently get the average change* in $F(x)$. This is simply the change that would be wrought in $F(x)$ by a unit's change in x under the average (as to x) of the conditions between a and b. You see that we have already passed into the hypothetical. Now by causing b to approach a as its limit, we narrow these average conditions down towards the conditions existing for a. Thus, the above change in $F(x)$ that a unit's change in x would induce, under the average of the conditions obtaining between a and b, is made to approach, as its limit, the change in $F(x)$ that a unit's change in x would induce under the conditions obtaining for a. You see that this is the identical process used in finding the derivative of $F(x)$, and so we see what a derivative really is. A change dx in x, under the same conditions as above, would evidently make dx times as great a change in $F(x)$ as a unit's change would make. Thus, to get what I have defined the differential of $F(x)$ to be, it is only necessary to multiply its derivative by dx. So we see that the ordinary process of differentiation does not lead to any real quantity, but to a hypothetical. Hence the reverse process of integration passes, not from any real but, from a hypothetical quantity. It is often necessary, therefore, to find a differential from the conditions of the problem.

---

*Notice that this is not a ratio, as is generally stated, but a concrete quantity of the same nature as (Fx).

To do this we have only to notice what conditions, in the unknown $F(x)$, correspond to some typical value a of x, and to determine the change that would result in $F(x)$ from a change, $d\,x$, in x if these were maintained. For example, if $F(x)$ is the arc of a curve, a plane area, or a volume, the conditions corresponding to any value of x are only direction, diameter, or cross-section, respectively. The well known differential of either is simply the change (due to $d\,x$) that would result in $F(x)$ if this condition were maintained. For a mass, the conditions are cross-section and densities pertaining to points in that section. For a planar moment or moment of inertia of an arc, plane area, volume, or mass, the conditions are the arm a (taking axes rectangular for brevity) together with the respective conditions just named. The last example brings in a fact of utmost importance where the position of the magnitude is one of the features of the problem. As the differential, corresponding to a, is the magnitude that would exist under conditions pertaining to a and to no other value of x, so that differential itself pertains to no other value of x but a, that is, this differential (supposing axes rectilinear) must be considered as lying in the plane $x=a$.

If we make x a function of the time, and thus put x and $F(x)$ in an actual state of change, their differentials, as defined above, become their fluxions. Thus, the fluctional calculus is only a special case of the more general calculus just proposed. In my opinion, the sole reason why the fluxional calculus is accounted weak is that the nature of its fluxions has never yet been set forth with proper emphasis. In the hands of one who emphasizes their hypothetical nature in his own mind, and who decides before applying it to a particular case what conditions he must suppose to be maintained, it becomes every whit as efficient as the infinitesimal calculus itself. And it is far ahead of the latter where actual change of any kind is to be dealt with.

I have already pointed out that the calculus, built on the foundation proposed above, is identical with the so-called infinitesimal calculus in whatever is essential, and that it differs from it only in being free from the irrelevant idea of infinitesimals, whose only function is to cover up a supposed mistake. I have just pointed out that this calculus includes the fluxional calculus as one of its special cases. It is readily seen, also, from what has already been said, that it likewise includes the limit calculus as a special case, the case in which $d\,x$ is supposed to be unity.

In returning the proof of the above paper to the printer I add a postscript.

The conditions (slope, diameter, arm, intensity, density, or what not) under which $F(x)$ exists, vary in general with x, but are definite for each value of x. Taking some value of x as a type, and referring to the bottom of page 105 and top of 107, it will be seen that the differential (being the combined effect that such of these typical conditions as are relevant to $F(x)$ would produce if they were to continue to operate without change through a specified hypothetical change in x really expresses these relevant typical conditions themselves. Moreover, the differential, being a function of x, applies to all parts of $F(x)$. Thus I now consider the differential of $F(x)$ to be an artifice, concise and general, for formulating the conditions which determine $F(x)$. And I would now define the calculus to be the branch of mathematics which connects a magnitude with the conditions which determine it.

This view of the matter seems to me to relieve the calculus of all trouble as to theory; and it certainly facilitates its use. The antagonistic idea, that integration is a process of summation, is easily shown to be a pure and untenable assumption.

THE TEACHING OF ENGINEERING, SPECIFICATIONS AND THE LAW OF CONTRACTS TO ENGINEERING STUDENTS.

By J. B. JOHNSON,

Professor of Civil Engineering, Washington University, St. Louis, Mo.

It is admitted that the drawing of engineering specifications and the writing of contracts is wisely left to engineers of experience and discretion and to their legal advisors, but is this a sufficient reason why the underlying legal principles of contracts and the elements of good practice in writing specifications should not be taught to students? Further, can a want of time be successfully offered as an excuse for a failure to include these subjects in the schedule of studies, even a four-years' course?

It is probably true that engineers fail as often in accomplishing the ends sought from imperfections in the specifications and in the contract as from faults in the design itself. Faults or mistakes in the specifications and in the contract are likely, also, to prove expensive blunders, and often lead to litigation and delay. If little or no mention is made of this department of an engineer's work the student is apt to regard it lightly and of little importance. He should at least be taught to respect it.

As to finding the necessary time for teaching this subject it is here as in many other matters—a few guiding principles well presented and a bird's-eye view of the entire field is all that is required to be given in the school. The young engineer can then be left to take his own way. From twenty to forty hours are quite sufficient for the whole subject.

What is most needed in this direction is a suitable text. The law of contracts, or such part of it as is required for the student of engineering, can be compressed into forty octavo pages. This is, furthermore, the simplest department of law and the one most readily understood by the layman. These forty pages may contain not only the statements of the legal principles involved but also such foot-note indicators of what not to do as may be requisite to guard the contract from many legal weaknesses.

The writer is fully convinced of the truth of the maxim that "when a man is his own lawyer he has a fool for a client," yet this applies with most force to the getting out of trouble. For keeping out of trouble some knowledge of the law is far more effective to the engineer than the constant privilege of consulting a lawyer at every turn which all Americans now fully enjoy!

The writer wishes to suggest a course in this subject which would become possible when a suitable text has been prepared, as follows:

I. A short synopsis of the Law of Contracts.

II. A description of the various papers, including

the Advertisement, the Instructions to Bidders, Forms of Proposals, Plans, Specifications, Contract and Bond, which go to make up the set of writings which constitute the complete contract between the parties.

III. A pretty full analysis of the nature of, and of the reasons for, a series of general clauses which would pertain to and might form a part of nearly all contracts let by engineers. In each instance the ordinary variations of the specification should be discussed and the conditions named under which such variations would be used; also some of the forms to be avoided and the reasons for the same with references to the legal principles given in Part I.

IV. A few sample specifications descriptive of certain elements of many kinds of engineering work, as for instance:

(*a*) Materials of Construction.

(*b*) Concrete.

(*c*) Masonry.

(*d*) Earthwork Excavation and Embankment.

(*e*) Earthen Dams.

(*f*) Stationary Engines.

(*g*) Pumping Machinery.

(*h*) Erection of Structures.

 Etc., etc.

It is to be understood that these would be elemental or fragmentary portions of specifications and would be given in the text as illustrations only. One complete set of papers, however, should be given to illustrate how all the documents pertaining to the letting of an engineering contract are related to and supplement each other.

This book should be an octavo volume of not more than two hundred and fifty pages, and should sell for about $2. To supplement this book there might be published in separate pamphlets, uniform in size for binding, complete model specifications for various kinds of engineering works and machinery, with blanks left for dimensions and other descriptions to adapt them for particular cases. These might be prepared by different engineers of experience and reputation, but they should all be published by the same house and all listed and priced in the book here described.

With such a text a very short course would suffice to place clearly in the mind of the young graduate the importance and spirit of this part of the engineer's duties, and he could not fail to realize that these few hours had proved as profitable and as fertile in results as any other similar length of time in his course.

The writer wishes here to express his conviction that we, as teachers, must do a great deal of sifting, and boiling down, on the one hand, and must also exercise much more care in utilizing the time of the student to the best possible advantage, on the other. We must more and more confine ourselves to the teaching of principles which are of service in practice, and to limit the teaching of methods to those only which illustrate and embody principles, and which lead to investigations and to demonstrations on the part of the student. That after the student enters his four-years' course of study in any department of engineering it is the mind and not the hand which must be trained, and that every moment must be zealously guarded and utilized. When this is done we will find that these four years will suffice for a far wider scope of subjects to be presented and probably for the acquirement of a more fruitful knowledge of these subjects than is now obtained. The immediate money value of the men who graduate from our schools is of less consequence than their ultimate power to invent, to advise, to direct, and to control.

MECHANICAL DRAWING IN TECHNICAL SCHOOLS.

By J. J. FLATHER,

Professor of Mechanical Engineering, Purdue University, Lafayette, Ind.

The comparatively recent rise of engineering as a profession and its rapid growth have called into existence numerous technical schools for the education of the future engineer which have been organized, and oftentimes controlled, by those who have derived their ideas of proper methods of professional training from inherited or acquired experience in teaching the old or "learned professions"—theology, medicine and law. In the "learned professions" intellectual development is the primary thought; whereas in the education of the

8

engineer the aim is to so train the student that he may have a thorough conception of the principles upon which the profession is based, and sufficient practice in applying these principles as will best enable him to succeed in whatever branch of engineering he may enter after leaving college. Whatever intellectual development the student may acquire is a result of the process, not the direct aim.

Considering the newness of engineering education it is not surprising that features exist, and are carried out in the curricula of some of our technical schools, which only in a measure attain the desired end. Owing to the comparatively short time they have been in use these features have not been generally recognized, or if recognized the adherents of the old school have opposed any change. Another reason for their existence is the fact that each new school is equipped with instructors and professors who usually adopt the methods they were taught; hence we see in certain lines a general similarity of instruction which, although it may promote the knowledge of the student and enlarge his mental capacity, does not best subserve his ultimate advancement.

While we recognize that the subject of engineering education is not sufficiently old to be crystalized into a definite form, and while we further recognize the advisability of having different kinds of engineering schools— some having departments for shop practice and others providing courses where manual training is omitted— yet there are many subjects which are commonly included in all courses for engineering students, prominent among which is mechanical drawing. This subject as frequently taught does not give the student that training, of which we have already spoken, which will enable him to make the greatest advancement after graduation. The actual system of projection as taught by many instructors is opposed to shop practice, and the student so taught, when he enters upon the duties of his profession, has not only to learn the shop methods but he must forget much of what he has previously learned.

As drawing is the language of the engineer it is of the greatest importance that the student be properly taught to express himself, and that his methods conform to those in vogue among practical men. You are doubtless familiar with Chordal's boy Joe who wrote twenty-two pages of legal cap to explain a detail drawing. A drawing which requires explanation does not fulfill its mission. A draughtsman can not always be at hand to interpret his drawing nor should he be. Drawings for use in the shop, right at the draughtsman's hand, should be so complete and self-explanatory that a properly skilled workman a thousand miles away could execute the work without a word of descriptive matter. The machine-shop drawing is simply a memorandum, showing what is to be produced. It is necessarily an illustrated memorandum, and to be perfect, it should answer all questions which a workman can reasonably ask in regard to the work. It is not necessary to have a finely finished drawing, in fact, shop drawings should

not be finely finished; a clear representation of the
piece with good heavy lines and distinct dimensions,
even if drawn freehand, is to be preferred for shop use
to a highly finished fine line drawing which the college
graduate is so prone to turn out when he first enters a
draughting room.

As freehand sketches enter largely into the work of
the engineer the importance of teaching freehand drawing can not be overestimated; and although it may not
be generally considered as mechanical drawing, yet its
common use in the drawing room and shop entitles it
to be properly classed under this general head. Moreover in this connection it is most frequently used to
give form to an object in a manner similar to that
employed in mechanical drawing. In this respect it
differs from artistic sketching which gives a general
representation of the prominent features of an object or
scene. Freehand drawing is a wonderful aid in the
cultivation of the observing powers, and, practically,
the first step in drawing is to learn to see accurately.
One of the foremost American engineers of the present
day, whose name you would recognize were it mentioned, has so cultivated this power of observation that
he has in several instances comprehended the details
and dimensions of a machine so completely, by inspection, that he has been able to make working drawings
and duplicate the machine without making a single
measurement or detail sketch.

The beginner will soon recognize that the untrained
eyesight is untrustworthy, and when this is realized,

the importance of keen observation becomes apparent. Erroneous conceptions which the untrained eye will cause the hand to execute may be corrected by intelligent practice, (and for this purpose the student should be early given a course in object drawing.) His first work should consist of exercises in drawing straight parallel equidistant lines, both horizontal and vertical; then should follow exercises with curved lines and circles in the flat, after which the student should be given a few examples in shading with the pencil from copy. After this he may be given simple objects and machines to sketch, and the relation of the several views of a mechanical drawing to the object represented, and to each other, should be early impressed upon his mind. For this purpose actual machine pieces and combinations will be found of greater value than geometrical forms or wooden models; for a machine piece not only presents a greater range in offering several different views which may differ in some respect, but it is a common experience that a boy takes to such drawing more readily and with greater interest. As the work progresses the student should take measurements from the objects to be presented and the sketch should be dimensioned according to the conventional shop methods. Another advantage in employing actual machine parts is thus obtained, for, where a measurement is a little above or below a standard size the student must exercise his judgment in deciding upon the suitable dimension to adopt—a point which the draughtsman must learn sooner or later. A valuable service in freehand drawing is given by introducing from time to time, memory sketches in which the student has to draw a given piece after an examination of the part. Time sketches are also an aid in teaching this work and should be alternated with the memory sketches.

An aid to the proper disposition of the views in a mechanical drawing is the conception (proposed by Mr. Sellers) that the object is surrounded by a glass case hinged on all sides, and so constructed that it can be opened out flat, as shown in the sketch on the blackboard. By assuming this case to be placed over the object to be drawn, the top view will be seen by looking on top of the object; the front view will be seen from the front; and the right and left end views to the right and left of the front view respectively. By tracing the outlines of these several views upon the glass and opening out the case the several views will be placed as follows: The top view will be shown on top; the front view will appear immediately under the top view; and the end views will be drawn to the right and left of the front view and in line with the latter, so that *the position of the drawing indicates the view taken.* This disposition of the views has so many practical advantges that it is employed almost universally in the shops.

Exercises in lettering should begin with the first work in freehand drawing and should be continued throughout the entire course. There is nothing which discounts a drawing so much as poor lettering, and there is scarcely one technical graduate in ten who upon leaving college can print his own name rapidly so it will look well. A part of this fault lies with the teacher who must bear constantly in mind that the old proverb, *"poeta nascitur non fit,"* does not apply to lettering. The man who would letter rapidly and neatly must acquire the ability by continued practice—he is seldom born with a pen in his hand. The student should be given one or two simple forms of letter which can be made rapidly, and all his subsequent work should be confined to these simple forms. It is folly to attempt fancy lettering and time spent on such work is thrown away, for no office draughtsman could spend the time necessary to make such letters even if desirable. A plain block letter for titles, and a running law italic, plain Roman, or a style of round writing, are the only forms which the student should try to master.

The writer has adopted a plan which is productive of good results and one which might be carried still further to advantage. It is that in all exercise work and in weekly tests, all subject headings, with name and date, must be neatly printed off hand. This gives continual practice in addition to that obtained in the drawing room, and it is found that the men gradually acquire a fair degree of rapidity and style.

In this ground work the principles of projection should be thoroughly inculcated in the mind of the student, as much of his after success and advancement depends upon his thorough knowledge of these elements. The time spent on this free-hand work is in many colleges too short to accomplish much good. It is our opinion that the average student is in a better condition to enter upon the more advanced work of drawing and machine design by a thorough course in freehand, and that, ultimately, he accomplishes much more than if such time were saved for actual instrumental work. From four to eight hours per week may be advantageously devoted to this work throughout the first term of Freshman year.

At this time instrumental work should be taken up, and right here, in the opinion of the writer, much valuable time is wasted in elementary exercises in line work, consisting of the construction of geometrical diagrams, arranged with a view to securing facility and accuracy in the use of instruments. Why not start the student upon projection drawing at once and let him acquire facility of execution at the same time he is obtaining practice in projection? His experience in free-hand sketching has taught him the principles of projection, and his subsequent work with the T-square and compass should be in the nature of a development.

Let the student then be given machine pieces and simple machines suited to the capacity of the individual from which he must make working drawings. By the use of notes, blackboard and elbow work let him know what constitutes a good working drawing and have him use the ordinary conventional shop methods of representation. Have good examples of shop drawings from leading manufacturers so disposed in the drawing room that he may refer to them for comparison and

for such minor details as are necessary to make a
proper drawing.   Above all let the student assume that
his drawing is going into the shop to be worked from
and treated in all respects with no more care than any
other working drawing.   If there is any error or mis-
understanding the draughtsman alone is responsible;
if the lines and figures become indistinct with shop use
it shows they should have been made heavier; if verbal
or written explanation (other than the customary notes
which appertain to a working drawing) are necessary
it shows an incomplete drawing; and it is the province
of the instructor to see that these objects are attained.
In this work the character of the examples may be
very properly differentiated for the several departments
of engineering.   In the present discussion of the subject we shall confine ourselves to such work as would
be suitable for mechanical or electrical engineers.

Tracing and blue printing should be taken up at
this time and the work of the term completed by a
series of graded exercises in isometic drawing, line
shading and tinting.   To carry this work through
successfully will require from six to ten hours per week
for twenty-four weeks.

At the beginning of sophomore year a short course
should be given the student in Descriptive Geometry
including shades and shadows.   The time devoted to
this subject may vary from five to eight hours per
week for sixteen weeks, two hours of this time being
given over to recitation room work.

Upon the conclusion of this subject the student
should be given a line of work preparatory to machine
design; for lack of a better name we shall call this
work the mechanism of machinery. Its object is to
study the nature, proportions, construction, and
analysis of existing machines, and may be taught by
means of lectures, inspection of the actual machine, if
available, and drawing room exercises. Where possible the machine should be taken apart by the students
and its several parts sketched. Visits of inspection to
manufacturing works in which the student may watch
and note the more important methods of construction,
constitute a part of this work. In the various machines
inspected the methods of oiling, and adjustment, provision for taking up wear, extent of wearing surfaces,
different metals used and reasons therefor, should, not
only, be pointed out to the student, but he should be
taught to observe these features himself. By noting
at the same time the widths of belt used and the speeds
at which the machines are run; also the character and
quantity of work turned out, our student is not only
preparing himself for the immediate work of machine
design, but he is acquiring habits of observation which
will be an advantage to him throughout his entire career.
The ability to get at the true inwardness of a machine
by observation and analysis from point to point is not
usually possessed by the untrained engineer. Lectures
on methods of handling and turning out work in
machine shops, and other works, constitute a valuable
feature of this subject, for, even with the best of college shops and methods of instruction, the student

obtains a limited knowledge of shop practice. The object of the lecturer in this subject should be to point out ways and means, and indicate where designs are defective; where they may be improved; where and why other proportions should be used; how parts may be changed so they may be constructed more cheaply; in short, as far as possible, in the limited time at his disposal, his object is to teach the more important principles which guide the successful engineer in his designing, as far as the constructive features are concerned; mere statements and facts should be subjected to the idea of bringing out the reasons *pro* and *con* in any given case.

Suitable exercises to be worked out in the drawing room should accompany this work. These may consist of combinations of several pieces shown in section and otherwise, so that the student learns the relation of the drawing to the thing, and in assembling these several pieces he learns the relation between the individual thing and the machine. For this work the exercises may be taken in part from sketches made by the student, and in part from blue prints of details of machine parts which may be obtained from the manufacturer or they may be taken from designs made by senior students. This work should extend through to the end of the sophomore year or for a period of twenty-four weeks. One lecture and from five to eight hours per week inspecting, sketching, and drawing may be very advantageously devoted to this subject.

The student is thus prepared at the beginning of junior year to take up the study of machine design. Machine design is both a science and an art. It is a science in that it is based upon rational deductions; it is an art in that it requires skill and judgment in applying these deductions to the engineering materials which are available. Calculation from a given formula is not machine design; judgment, or what has been termed "horse sense" enters quite prominently as a factor in the work. If calculation calls for a sixty inch boiler shell to be two inches thick for one hundred pounds pressure the designer must know intuitively that something is wrong. He should also be sufficiently informed on shop practice and methods, that when he designs a piece of work he must know whether it can be made in the shop. As Reuleaux states, in the designing of machines and mechanical constructions, the draughtsman must draw from his knowledge of well-known forms and parts and combine them; but to proportion them properly, and adapt them to the purposes required he must take into consideration the stresses to which they are subjected, and the behavior of the available materials under varying conditions. The choice of suitable dimensions and forms is the work of the machine designer.

Engineers and practical shop men are enabled to proportion the various machines which are daily manufactured, and to satisfactorily design new machines without a knowledge of the character of the forces acting in the various parts of the machine; but it is safe to assume that this knowledge has been obtained only as a result of long experience and the process of substitution—that is when a part failed in a machine a larger or stronger piece was substituted in the next one built—an expensive method of design. The shop designer is frequently an averager who obtains his results largely from the catalogues and publications of other manufacturers; and while this is valuable as affording a means of comparison and corroboration, we believe that all intelligent mechanical design must be based on a knowledge of scientific principles. On the other hand there is difficulty in proportioning many machine parts on account of the lack of exact information, and of their variable functions not amenable to exact rules. For such cases empirical formulas and tables made up from the results of experiment and practice will be found of great assistance to the designer. From these considerations it will be seen that the province of the instructor in this subject is to teach the principles available as guides in machine design, and to furnish such examples to be worked out at the drawing-board as will ultimately tend to the development of the student's engineering knowledge.

The teacher in machine design should evidently be a man familiar with practical shop work and modern methods, but he should not attempt to cram the student full of facts and tables. If the underlying principles are made prominent let the instructor feel that his work is well done; for development into the full fledged engineer can not occur at once. It takes time, and we must remember that there is inertia in mind as well as in matter. The teacher of machine design must also guard himself against giving the student formulas without showing their derivation and the relation of the several factors involved. It is not teaching principles if we select a dozen or more formulas with constants already determined and say to the student draw a piston rod end, for instance, making the diameter of rod one-sixth the diameter of cylinder. If this is the way to teach our boys why not get out tables for general cases and put them into the hands of our students? The average student is a reasonable and a reasoning person, but even a student is limited in capacity. For this reason we should so much the more teach principles as foundations and let the character of the individual problems be determined by the ability of the man. It is unwise to gauge all men by the same standard, and the old saying "You can't make a whistle out of a pig's tail" should constantly be borne in mind by the college professor who puts all his men through the same work in design. If a boy has capacity for engineering he may be given problems and designs to work out that a man of less natural ability could not possibly execute. Although natural aptitude is a great advantage in this work, much can be acquired, and the teacher must judiciously direct the work of each student in order that he may thereby the better prepare the individual for his future vocation.

Before entering upon the real work of the drawing room, it is advisable to spend about three weeks in the recitation room (three hours per week) in a general discussion of the engineering material; and the principles and general rules of calculation, which are applicable to problems that may arise in the determination of machine parts. The subject matter taken up under these heads is not intended to displace, nor should it be substituted for, the study of Metallurgy and Applied Mechanics. The properties and strength of engineering materials, and the principles relating thereto, are to be considered as an equipment of suitable tools which will enable a man to intelligently design those simple elements of which a machine is composed. The hasty preliminary view of the field is merely to indicate to the student where these tools may be found.

From six to ten hours per week for twenty-eight weeks may be devoted to the design of machine elements such as those which come under the head of fastenings, gearing, pulleys, shafts and bearings. This may be followed by problems in Kinematics relating to straight line motions, quick return motions, and various cam motions, after which practice is given in the applications of Zeuner's diagram to valve gears. Eight to ten hours per week for twelve weeks are necessary for this portion of the work.

At the beginning of senior year steam engine design may be taken up and carried on for sixteen weeks, seven to nine hours per week being necessary for the work. Upon its completion the student takes up the design of machine tools, boilers, pumps,. clutches, hoisting, punching, and other machinery. The time allotted to this section of the subject should vary from eight to twelve hours per week for twenty-four weeks. It will be found that quite a number of men would even spend as much as twenty hours per week at this work, if permitted to do so.

The work in machine design as carried out by the writer at Purdue University, varies with its character. As far as possible the students are given individual problems so that each may make his own calculations, and become familiar with the method and application of the formulæ involved. Preceding each new series of problems a short lecture is given to the class in which the underlying principles relating to the problems are pointed out, and the derivation and application of the formulæ to be used are explained, and reasons noted why other proportions should be adopted under other conditions. In these problems each part is designed to do some definite work for which the data is given, or may be determined. Upon the completion of these elementary problems, the student is given a course of exercises in Kinematic and valve gear problems as previously indicated.

With the commencement of the senior year he takes up steam engine design. In this work a given horse-power of engine is assumed, suitable limits are chosen and indicator cards drawn for minimum, maximum, and normal conditions; these diagrams are modified to conform to practical conditions; and the diameter, length

of stroke and speed of the engine are determined after
the tangential pressures upon the crank have been
ascertained from assumed inertia effects. Instruction
is given by means of lectures and drawing room work.
Each student makes a complete set of theoretical diagrams in pencil on sheets 24x36'' (the standard size
for senior drawings) after which the details of the
engine are worked up by the class collectively—one
man takes the cylinder, another the bed, another the
cross-head and connecting rod, and so on. In designing these details the student has access to a good
reference library in the drawing room and a very complete collection of trade catalogues and working blue
prints from manufacturers, so that he has every
means of becoming informed in regard to current
practice.

In the design of details and in machine drawings
which follow this work the tracings are made direct
from the pencil drawings. When completed each student obtains a full set of blue prints. Conventional
shop practice is used throughout all the work in
machine design, and this is especially emphasized during the senior year where greater occasion for its use
arises. For calculation and preliminary notes, a system has been adopted by which the work and results
are kept distinct, and yet both are preserved for future
reference or comparison. The advantage of any such
system in locating errors is apparent. In the present
case, notes, formulæ and results are kept on one page
while the calculations are made on the opposite page,

9

so that the work and results are together, yet distinct. No attempt is made to economize paper—the only object aimed at being a saving of time, and this is best attained by teaching the student to systematize his work and have the calculations where they may be referred to in case of error or in checking results. During the progress of the work, especially in scale drawings, the draughtsman is encouraged to make full-sized sketches of the several parts of his design in order that he may obtain a better degree or proportion. To this end specially prepared section paper is provided so that the part may be rapidly sketched in free-hand without other reference to size than that furnished by the paper itself.

The concluding work of the senior year consists ef special designs, assigned individually according to ability and circumstances. This work is generally presented in the form of a problem which involves more or less engineering knowledge, and the exercise of judgment in the application of principles studied earlier in the course. Time blanks are used for these designs, as it is thought that a record of the work and time spent on it is advantageous to both student and instructor. Instruction is given in this work by means of consultation with the student, advice, and corrections. References are cited when the problem is assigned. As the design progresses many of the details are sketched off and worked up by under-classmen. For this purpose several of the more advanced Freshmen and Sophomores are detailed as assistants to the Seniors, and from the experience of the last three years, the plan can be endorsed as producing excellent results. The Senior is given a responsibility which insures greater care and thoroughness on his part; besides which, it enables him to accomplish a great deal more of real shop work than would otherwise be the case. There is also an advantage to the under-classman as he is brought more intimately in contact with a class of work, as nearly as possible, similar to that carried on in engineering drawing rooms.

Recapitulating and taking the average time allowance, it will be seen that we advocate a course in mechanical drawing covering four years and occupying a total of one thousand, one hundred and sixty-eight hours allotted as follows:

		Hours per week.	No. of weeks.	Total hours
FRESHMAN.	Free hand drawing and lettering.	6	16	96
	Instrumental drawing; Tracing; Blue-printing; Shading; Tinting; Isometric.	8	24	192
SOPHOMORE.	Descriptive Geometry; Shades and Shadows;	6	16	96
	Mechanism sketches; Assembling and section drawings.	6	24	144
JUNIOR.	Design of machine Elements; Cam motions and Valve Gears.	8	40	320
SENIOR.	Steam Engine Design.	8	16	128
	Machinery Design.	8	24	192
	Total			1168

SOME GERMAN TECHNICAL SCHOOLS.

By STORM BULL,

Professor of Steam Engineering, University of Wisconsin, Madison, Wisconsin.

The paper which Professor Swain read before Section E. of the Engineering Congress in Chicago last year, "Comparison between American and European Methods in Engineering Education," created quite a little interest; and as it is only two years ago that the writer went purposely to Europe to visit the polytechnic schools in Germany and Switzerland, it was, therefore, thought that he possibly might add something of interest to this discussion.

The writer is more familiar with the Polytechnikum at Zurich than with any other school, having studied there four years, from 1873 to 1877. During the visit to Europe in 1892, he visited the schools at Hanover, Aachen, Carlsruhe, Zurich, Munich, Dresden and Berlin, and it is of these schools he shall speak in what follows. The writer had hoped that he should have ample time to prepare a thorough going paper on the subject, but some work for the University of Wisconsin during the present vacation has required his constant attention, and he is afraid that his paper will not be what the members of the Society have the right to expect.

In comparing the schools mentioned above, it is absolutely necessary to keep in mind that these schools have greatly changed during the last ten or twenty years. There was a time when the school at Hanover was considered the best school of all. Again there was a time when Carlsruhe was supposed to be at the head of the list, and Aachen certainly had for a while as good a reputation as any. Now, these three schools, especially the first and the last mentioned, are very far below the others both as to equipment and teaching forces. The principal reason for this change is that Prussia tries to concentrate everything at Berlin, and necessarily then, Hanover and Aachen, both in Prussia, must suffer. Some fifteen or twenty years ago the school at Berlin was far from being the best; now, on the other hand, there can hardly be any doubt about this school taking the first place in Germany, both as to equipment and as to teaching force. The new buildings of the Polytechnic school at Berlin are very fine, so fine, indeed, that some of our western legislators, who hold the purse strings for the state universities, would be positively frightened if they should see them. However, it should be stated that the buidings of all the schools mentioned, except that of Carlsruhe, are very good, and usually fine looking, and I have thought that for most of these schools it would have paid better to spend less money on the buildings and more on the equipment of laboratories. The school at Berlin has had ample means, so that it has been able to pay its professors higher salaries than elsewhere; it

is, therefore, but natural that, taking it as a whole, the
school at Berlin possesses a stronger teaching force
than the other schools. The number of students has
been increasing, and I find that during the winter
semester of 1891–92 the number of regular students
was 1,380; and of "hospitants," 511; in all, therefore, 1891, of which more than 800 then belonged to
the department of mechanical engineering.

I have already stated that the schools of Hanover
and Aachen do not occupy the rank which they once
had. Of course they have not retrograded, but they
have not been able to keep pace with the rapid progress of the other schools. The number of students
has, therefore, diminished, and these schools will
probably only occupy a secondary position in the
future, especially as I was told that the Prussian government draws all the best teachers from these schools
to Berlin as soon as there is a vacancy. The three
schools at Dresden, Munich and Karlsruhe occupy
relatively to the others about the same position. They
are all three at the capitals of the states of Sax-
ony, Bavaria and Baden, and it is the pride of these
countries to have a good polytechnic school, independent of the school at Berlin. We find, therefore, that
these schools have made notable progress during late
years, and that they have been able to keep some of
the most famous technical educators among their professors. That they have not been able to keep up
with Berlin is true enough, still these schools are well
equipped, and they have an ample teaching force.

The number of students is, of course, small compared
with that of the school at Berlin. They have been
able to retain some of the most famous teachers and
investigators until this day. I mention: Dr. Gustave
Leuner, in Dresden; Professor Bauschinger in Munich,
and Dr. Grashof in Carlsruhe, of which, however, if
my memory does not deceive me, the two last ones
have died within the last year or two. The number of
students at Dresden were, in 1891–1892, 394; in
Munich, 891; in Carlsruhe, 587. Comparing these
numbers with the 1,891 students at Berlin, the schools
seem rather small. In a class for itself stands Zurich,
which at all times since its foundation in 1854, has
stood well in the front rank of polytechnic schools.
The little republic of three millions or less of inhabi-
tants has shown itself very liberal in regard to this
school, so that at the present time no other school possesses such large and well adapted buildings as Zurich.
I will here only mention the new chemical and physical laboratory buildings. There are probably some
universities in the world which have as good a chemical laboratory, but I am very certain that there is
nothing like the physical laboratory in Zurich, elsewhere. However, it has been found difficult to keep
their best men there, especially as so many were Germans, who preferred to stay in Germany instead of in
Switzerland; the reason being that the salaries of all
officers of the government of Switzerland are very low,
and, as the Polytechnikum is a federal institution, it
has been found difficult to get the consent to give the

high salaries demanded. Of the professors which Zurich has lost in this way, I may mention: Reuleaux, Zeuner, Clausius, but the school has been able to retain such men as Cullman, Fliegner, Ritter, etc. The number of students was, in 1891–92, 1,089, of which about one-half were foreigners. It is very noticeable that both the absolute and relative number of students from foreign countries is very much larger than in any other school, and this, notwithstanding the fact that it is not so easy to gain entrance there as in the others, as they are required to pass a very rigid entrance examination in Zurich, whereas a testimonial from a preparatory school is sufficient in Germany. It is, therefore, a well known fact that a goodly number of those who are rejected in Zurich at entrance, and also later during their course, turn to Dresden and Munich, and are received there without examination. The high popularity of the Polytechnikum is then not due to the easy entrance nor to the superior professors—because others certainly have as good—nor to cheap living, because it costs more to live in Zurich than in Germany, but to the good equipment, and especially to the method of instruction and to the requirement of the student in Zurich. I might possibly also add to these last named reasons that of the courses of study in the various engineering departments, which courses seem to be very excellent.

I have now come to a point which brings me in direct connection to the paper read a year ago by Professor Swain, the methods of instruction in German polytechnic schools. Evidently Professor Swain has not included the Zurich Polytechnikum in that number of German schools, from which he characterizes the German method or instruction, and that is, perhaps, natural enough, as Zurich is not a German polytechnic school, although often included among them, and because the method of instruction at Zurich is, in more than one point, radically different from that in vogue in German polytechnic schools. In all German polytechnic schools the instruction is given by means of lectures exclusively. In Zurich there is no study in which part of the instruction is not given by lectures, but at the same time there are no required studies in which there is not a weekly or bi-weekly reptitorium, as it is called there, where the ground covered by the lectures of the previous one or two last weeks are gone over. These review classes are conducted partly by the professor, partly by his assistants; and the large classes are divided into groups so that there are never more than twenty in each group. Here the students must recite and very few absences are tolerated from these reviews. The roll is called and absences noted. If a student should stay away from say two of his recitations without good reason, he will be called before the dean of the department and given a reprimand; and if it occurs again he will have to appear before the director of the Polytechnikum, and receive a very severe reprimand, in which he is told that, unless he changes his tactics, he will be required to leave the

school. It occurred several times during my stay at
Zurich that students were dismissed from the school
because of inattention to these matters. It goes without saying that a student who attends the reviews will
also be dismissed if he shows that he is not learning
anything. This might not occur in a very unimpor-
tant branch, as the professor in such study probably
would not push it so far. Failure or inattention in
such study would, however, be elements in determining the status of a student in the school, if he should
have proved a failure in something else. The students
at Zurich are required to come to the lectures; however, as no roll is called, anybody may stay away without any record being made of it. But as they must
come to the reviews, and as they must show that they
know something, the result is that the attendance at
the lectures is very much better than it otherwise would
be. In fact, it is altogether an exception that a student
cuts a lecture at Zurich. They find it is the easiest
way to learn something about the subject to go to the
lecture. Books of reference, which nearly always are
given, may, of course, help a great deal, but it is the
exception that the lecturer treats the subject in exactly
the same manner as found in the books of reference.
It is, therefore, my experience from Zurich that,
owing to the regulations in vogue there, the attendance
at the lectures is all that can be desired. At the same
time a good lecturer will have a better attendance than
a poor one. In passing I will state that it is my opinion that it is absolutely necessary that a professor

should be a good lecturer if the lecture system shall be a success.

The lecture system as practiced in Germany is, then, very different from that in Zurich. By omitting these required reviews the personal contact of student and professor is lost, except where there is work in the draughting room or in the laboratory, and the amount of work which the average student does is greatly reduced. That this last assertion is a fact, I think that everyone at all familiar with the schools at Zurich and Germany will admit. I was not reported correctly in the discussion of Professor Swain's paper published in the proceedings of this Society. I am there made to say that Professor Schroeter, in Munich, had inaugurated the methods of the Zurich school at Munich. What I did say was that Professor Schroeter expressed the strongest wish to do it. At the same time he told me that it was out of the question, as the German students insist strenuously on what they call "*Lernfreiheit*," liberty to learn or not as they please. Professor Schroeter, who had been an assistant in machine design at Zurich before he came to Munich, told me that, from his experience, there could not be any doubt that the majority of students worked much harder in Zurich than in Munich. The really studious ones will work equally hard, whether there are restrictions or not, but a great many young men at the age of from eighteen to twenty-five will be very liable to forget that they have come to study, and not to have a good time. Some few years ago the society of graduates

from the Zurich Polytechnikum petitioned the Schulrath—Board of Trustees—to abolish the requirements
of attendance at review classes. The answer of the
Schulrath is a very lengthy one, and is published in
pamphlet form. It gives the very best reasons for
refusing to change the policy of the institution, and I
think it is an admirable answer to the boasted advantages of the "*Lernfreiheit*" in Germany.

I agree completely with Professor Swain that the
preparation of the students in Germany or Europe and
in the United States is so different that because the
lecture system is or may be a success over there it may
not be here. That the students entering the schools in
Germany at Zurich are very much further advanced
than the students entering our technical schools, there
is no question. I am of the opinion that they are about
two years ahead over there. Now the lecture system, if it
is adapted to the entering student in Germany, should
be also adapted to the juniors in our colleges. And I
am of the opinion that this is the case. It is, perhaps,
not the place here to enter into a discussion as to the
relative merit of the lecture and of the recitation, as a
method of instruction. However, I must state that,
judging from my own experience in Zurich, the method
used there, of combining the lectures with a weekly
review, worked admirably. I have never been able to
see the great advantage in having the students daily
get up and recite something which they have learned
by heart for the occasion. And this will inevitably be
the case in a great many classes, depending, of course,

somewhat on the subject taught. On the other hand, the weekly or bi-weekly review will be something like an examination, where, in general, no committing to memory will do any good, and because, no "pony" is in use in Europe, as far as I know, the student will have to rely on his own knowledge acquired by hard study.

The notes which the students used to take down at the lectures were, in general, very serviceable, and I know from my own experience that these notes have been of a great deal of use to me after graduation. It is, however, true that some students are unable to take down notes in such a form that they will be of any value to them afterwards; and for these students, the notes furnished by the professor, or gotten out by the students themselves, furnish all the means necessary for acquiring a complete knowledge of the subject in connection with the attendance at the lectures. I take it for granted that a verbal explanation of a certain difficult matter is very much superior to a reference to a text or reference book. It is very likely true that a student may learn to understand it all by studying the book. However, often this is not the case, and the teacher must explain the matter in the recitation hour. This is what the German professor does when he is lecturing, and I am sure that because of the lecture the student will learn the subject of the lecture more easily than if he had had to study it out of a text-book. It puts more work on the professor, but lessens it for the many students. The disciplinary value of a recitation I do not underestimate, but for students at a university it should not be necessary to have these disciplinary exercises every time a study is taken up, but it should be sufficient to limit it to once a week or once in two weeks, as practiced at the Polytechnikum of Zurich. As a matter of fact, I know that we got over more ground there than here in the same amount of time, and I know, too, that, notwithstanding the fact that in general people work less in Europe than in the United States, yet the students, as a class, did not spend any less time there than here in their studies, rather more, the result being that at graduation the students really had a good knowledge of the studies which they were supposed to know. There is, of course, the same difference there as here between poor and good students, but the Zurich system makes it impossible for students to stay any number of years at the Polytechnikum pretending to study to become an engineer. In fact there is a rule, which is strictly enforced, that a student can not stay in any one class more than two years.

The personal relations between the teacher and the students, which I have alluded to before, are, I think, fully as close at Zurich as here in the United States. It is true enough that the professor leaves the room when the lecture is over, but at the same time it is also true that the students very frequently go to the professors in their private rooms for explanation of some obscure point, and, as far as I know, these explanations were always freely and willingly given. Again,

when the students get out notes for the lectures by
some multiplying process or other, the professor is
always willing to help them all he can. I have here
been speaking of Zurich. In Germany things are rather
different, and Professor Swains' descriptions of the
relations between students and professors are probably
correct. However, work in the laboratories or in the
draughting room, where the direct contact of the teacher
and student must take place, is so frequently connected
with the lectures, that the objection on this account to
the lecture system has not so much weight as might at
first sight be supposed. It will, of course, be objected
that I am prejudiced in favor of the lecture system
because of my studying at Zurich. At the same time
I have now taught in the United States for fifteen
years, and mostly by means of recitations, so that I
suppose that I am sufficiently informed to draw my
conclusions. I repeat again, however, that I do not
advocate the lecture system for freshmen and sophomores, only for juniors and seniors. I might mention
another advantage of this system, and that is, that in
purely technical subjects the matter must necessarily
change from year to year, and that can not be done
when you use a text book and have recitations exclusively. If you desire to introduce some new matter,
you must get out notes especially for the purpose, and
I have found that it is very confusing to mix the notes
and the text book. If the professor can furnish notes
of his lectures in advance, if he has a weekly recitation
or review, I do not see that any serious objection can

be raised against the system, and I can see quite a number of advantages. I have dwelt a long time on this subject because I think it a very important one, and because I know the prejudices of most American professors against the lecture system. If I have been able to remove some of the prejudices from some of you I shall feel more than gratified.

Returning to the subject of German polytechnic schools, I would like to say a few words about the equipment of these schools. I will first mention the draughting rooms. As far as I know all schools are provided with so many and large draughting rooms that each student has a desk exclusively for himself. I know it is the practice in some American technical schools to assign the same desk to students of different classes. I am very certain that this ought only to be done when it is absolutely necessary, and only for a short time. From my experience in Zurich and in Madison—especially the first place—I know that a double assignment would have interfered seriously with the work of the student. We always used to work a good deal outside of the regularly assigned hours, in the draughting room, and such work, which certainly ought to be encouraged, would have suffered greatly by having two students use the same desk.

The students in mechanical engineering in Germany are not required to work in the machine shops. There is generally a small shop, and the students have the privilege of working in it by paying a fee. The facilities are, however, small, and I am of the opinion that this nonrequirement is a serious want in the course. Of course, students are recommended to work in a machine shop, either before coming to the school or after leaving it. For those who have worked in this way before coming it is all right, but for the others, and these are the large majority, the lack of practical work certainly reduces the value of the instruction in several studies. The same thing might be said in reference to the lack of laboratory work in testing of materials. Most of the schools have a testing machine, and it is used for exhibition; some specimen is tested and broken in the presence of the students, otherwise it is only used for investigations by the professor. An exception should here be made in favor of the school at Munich, where Bauschinger's laboratory, which is very fine, is used also by the students, I have been told. The same story should be told about hydraulic and steam engineering laboratories; they have very little over in Germany in this direction. Whatever apparatus there are, are used for exhibition to the students and for investigations by the professor. Alone Munich, again, has a steam engineering laboratory, which is used regularly by the students in making tests of engines and boilers, etc. However, even there the equipment is so small that we would not be satisfied with it at all in the United States. In Berlin they have a gas engine, which, I understand, some few select students use for testing, but there it no regularly organized laboratory work. With several of the schools, notably Berlin, Munich and Zurich, are con-

10

nected experiment stations for testing all kinds of material, both chemically and mechanically. These are superbly equipped, but they are not used for instructional purposes except for an occasional visit to them by the students under the guidance of the professor. I understand, however, that a change in that respect is going to take place in Zurich, where but recently the station was established, and which station has ample facilities as a laboratory for students.

The laboratory work in physics has not been required of technical students until in recent years in Germany, and it is not yet required in all of them; in fact there are no facilities in several of them. But in others again the work is pushed with a great deal of vigor. Notably in Zurich, where the Physik Gebaeude contains laboratories so well arranged and equipment in such abundance that it stands unequalled at the present day.

I need not mention that as to chemical laboratories the schools in Germany and Zurich have all that can be desired. This is the more necessary because all of the schools graduate chemical engineers. It is a well known fact that the chemical industry stands on a very high level in Germany and Switzerland, and I take it that this is, to a great extent, due to the chemical engineers graduated by these schools. I know that some schools have a chemical course, but at the same time, the courses offered here are very incomplete compared with that offered for instance in Zurich. Would it not be well if we paid more attention to this matter here in this country?

The collections of models and specimens are usually very large in the German schools, in fact, very much greater than found in any school in this country. Take, for instance, the Reuleaux collection of Kinematic models or Dr. Wiener's models for descriptive geometry in Karlsruhe, or the technological collection in Zurich. These are very valuable for instructional purposes, but I think, however, a good laboratory equipment is still more necessary. Having mentioned the technological collection at Zurich it reminds me of the fact that there are very few of the American schools that give instruction in what the Germans call "Mechanical Technology," or in other words, in the various branches of industrial activity. This is, of course, due to the fact that the American schools have not so far felt able to let their professors specialize so much as is done in Germany, where, I suppose, there are, on the average, at least two professors for each branch to one here. Professor Swain has brought this out in his paper so fully that I shall not touch upon it any further.

I do not think I have the same impression as Professor Swain as to the lack of sufficient assistants. There is no doubt that the number of these assistants is very much smaller in Germany, compared with that of the professors here. But I think that is explained by the fact that there is so much less laboratory work there than here, and it is especially in such work that the assistants are required. On the other hand we find that these schools swarm with various kinds of mechanics who help the professors in various ways, for instance, in getting ready for the lectures when experiments are to be shown; or they are making apparatus for the various departments.

I have no doubt that this paper has been found long enough before now, and, although there are several points yet which I would like to mention, I will conclude, and in so doing I want to ask the leniency of the audience, both for my English, which certainly is not faultless, and for any other lack which might be due to the hurry in which, owing to circum stances, I have had to prepare this paper.

THE ORGANIZATION AND CONDUCT OF ENGINEERING LABORATORIES.

By GAETANO LANZA,

Professor of Applied Mechanics, Massachusetts Institute of Technology, Boston, Mass.

The objects to be accomplished by the Engineering Laboratories of Technological schools should be as follows: *First*, to give the students practice in such experimental work as engineers, in the pursuit of their profession, are called upon to perform; and, at the same time to make them understand better the bearing of their class-room work upon such experimental work. *Second*, to afford the students some experience in carrying on original investigations in engineering subjects, with such care and accuracy as to render the results of real value to the engineering community. *Third*, by publishing, from time to time, the results of such investigations, to add gradually to the common stock of knowledge.

In considering, not only the equipment, but also the organization, and the conduct of such laboratories, we should first regard the functions they are to fulfill as a part of the course of instruction of the students; for it is primarily for purposes of education that they exist in a school.

Their special equipment, in any one school, must depend upon circumstances, and especially upon the means at command; and the problem that generally presents itself is how to spend a given limited sum of money to the best advantage.

No general rules can be formulated for the solution of the problem, and each man who has charge of the equipment of such laboratories must exercise his own judgment with his own special circumstances in view.

Nevertheless, the following principles are applicable in all cases.

First. The equipment should be of such a character that the work required of the students may be real engineering work, and not mere playing with toys. Hence the laboratory should not be equipped with machinery of such small size as to be unsuitable to use in practice, nor should it be of such a character as to render it necessary to use it under conditions essentially different from those which occur in practice.

Second. The equipment should be such that it shall be possible to require careful and accurate work of the student, and not to tolerate careless or slipshod work.

Third. In choosing the equipment there should be taken into consideration the work being done, or to be done in the class-room instruction, in the same institution, so that the laboratory may serve the purpose of making the student have a more thorough grasp of what he is taught in the class room.

Fourth. It is a good plan to select the apparatus in such a way as to afford as much variety as possible, and also to give opportunity to do the same thing by different methods, one furnishing a check upon another; for instance, in some hydraulic experiments it may be well to determine the quantity of water delivered by two or three different methods.

Outside of these, and possibly a few other general principles which I may have failed to mention no general rules can be laid down, and each one who has the equipment to decide upon will have his own special ideas, and his own special objects to accomplish.

I shall, therefore, merely give a brief description of the equipment of the engineering laboratories of the Massachusetts Institute of Technology, and shall then proceed to discuss the proper organization and conduct of such laboratories, this discussion being the principal object for which this paper is written. The engineering laboratories of the Massachusetts Institute of Technology may be classified as follows, viz.:

First. The laboratory for testing the strength of materials, otherwise called the laboratory of Applied Mechanics.

Second. The Hydraulic Laboratory.

Third. The Steam Laboratory.

Fourth. An undefined remnant which may be called the rest of the Mechanical Engineering laboratory.

These laboratories occupy two floors 50x150 feet each. The laboratory of Applied Mechanics is furnished with the following apparatus: A horizontal testing machine of 300,000 pounds capacity, and of the Albert H. Emery type, suitable for testing a compres-

sion specimen eighteen feet long or a tension
specimen twelve feet long, furnished with the
necessary holders, platforms and grips, for testing
tensile or compressive specimens or wide riveted joints;
and with a suitable crane for lifting the specimens,
holders, etc.; a testing machine of 100,000 pounds
capacity for determining the transverse strength and
stiffness of beams up to twenty-six feet in length, this
machine being used, not only for tests upon wooden
and iron beams subjected to concentrated or distributed
loads, but also for tests of framing joints, and of the
riveted work in riveted iron girders; a testing·machine
of 18,000 pounds capacity for determining the transverse
strength and stiffness of beams up to fourteen feet in
length; a testing machine of fifty thousand pounds
capacity, for determining tensile and compressive
strength, and elasticity; a machine for testing the
tensional strength and stiffness of shafting, up to three
inches in diameter, and twenty feet in length; all the
necessary machinery and apparatus for testing the
strength (of whatever kind) of mortars and cements;
machinery for measuring the angle of twist of shafting;
for testing the tensile strength of ropes, with different
modes of fastening, and with different hitches; for
testing the effect of repeated stresses upon the elasticity
and strength of iron and steel; for determining the
strength and elasticity of wire; for testing the strength
of pipe, and pipe fittings under hydraulic pressure;
also accessory apparatus for measuring stretch, deflection, and twist.

The Hydraulic laboratory contains a closed steel tank five feet in diameter and over twenty-seven feet high, arranged for the insertion of orifices, mouth-pieces, and other special pieces of apparatus, with gates for controlling the discharge, and with connections for supplying water, in experiments upon pipes and motors. This tank is connected with a ten inch stand pipe over seventy feet high, so arranged that a constant head may be maintained, at any desired level. A steel tank of about two hundred and eighty cubic feet capacity gives opportunity for the measurement of larger quantities of water than can be weighed directly during the experiments. A system of pipes fitted for the insertion of diaphragms, branches, and other apparatus for studying loss of head, and the laws of discharge. An attachment has been fitted to the main tank, containing a Pitot tube, for studying the laws of velocity in jets, and for measuring the cross sections of jets. There are also two three inch water meters, and others of smaller size, and a variety of mercury gauges, weirs, standard orifices, mouth-pieces, diaphragms, nozzles, etc.; a six inch turbine arranged to be run under various conditions, a Pelton water motor four feet in diameter, so arranged that its efficiency can be determined under different conditions; other smaller motors, and a hydraulic ram.

The Steam Laboratory contains a triple expansion engine with cylinders of nine, sixteen, and twenty-four inches diameter respectively, and thirty inches stroke, arranged in such a way as to be run single, compound, or triple, as desired for purposes of experiment. This engine is of the Corliss type, and has a capacity of about one hundred and fifty horse power when running triple, with an initial pressure of one hundred and fifty pounds in the high pressure cylinder. It is connected with a surface condenser, and the other apparatus necessary to adapt it to the purposes of experiment. It should be observed that an important requisite for such an experimental engine is size of cylinder, for an engine where the cylinders are small does not operate with the economy of one where they are large, and hence investigations made by the use of small engines which consume thirty or forty pounds of water per horse power per hour, do not give results applicable to large engines which consume from twelve to fifteen pounds of water per horse power per hour. This laboratory also contains a sixteen horse power engine, and an eight horse power engine for giving instruction in valve setting, etc. It is equipped with several surface condensers, steam pumps, injectors, and ejectors, calorimeters, mercurial pressure and vacuum columns; apparatus for determining the quantity of steam issuing from a given orifice, or through a short tube, under a given difference of pressure; apparatus for testing injectors; and with indicators, planimeters, gauges, thermometers, anemometers, and other accessory apparatus.

The engineering laboratories are provided with a number of friction brakes; with machinery for determining the tension required in a belt or rope to enable it to carry a given power at a given speed with no more than a given amount of slip; with three transmission dynamometers; with a complete set of Westinghouse air brake apparatus, including the parts belonging to the car and to the locomotive, with the pump and valve of the New York air brake, with several weirs for measuring water, also two orifice tanks for the same purpose; with an experimental steam engine governor, with three machines for determining the coefficient of friction of lubricating oils; with a gas engine; with a locomotive link model; with a centrifugal pump, a gang pump, and a rotary pump; with a plunger pump, with a hot air engine, with a pulsometer; and with cotton machinery as follows: Two cards, a drawing frame, a speeder, a fly frame, a ring frame, and a mule, as well as accessory apparatus. There are available for experiment also two sectional and four horizontal tubular boilers; also a forty horse power engine, a number of looms as well as other apparatus.

The organization and conduct of the engineering laboratories of any given school should be of such a nature as to be in accord with the following principle, viz.: The prime object in view should be the best effect upon the advancement of the students, and no other considerations should be allowed to interfere with this. The first two functions to be fulfilled by these laboratories in the course of instruction are, as was stated at the beginning of this paper, to give the students practice in real engineering experimental work; and to make them understand better the bearing of their class-room work upon such experimental work.

Hence any organization which aims to make the laboratory independent of the regular class-room work, or which seeks to bring about any other than the most intimate relationship between the two is not for the best interests of the students, and should not exist. The grave disadvantages of any such scheme will manifest themselves in various ways, some of which are the following:

First. If the laboratory is operated as an independent course, the time consumed in explaining the nature of the apparatus, and the nature of the test to be made, as well as the manner of working up results, will be very great as compared with the time required if the laboratory work is co-ordinated with the class-room work.

Second. The instructor will have to explain to the students how to work up the results for each special test, giving him rules for such working up, instead of requiring him to find the solution himself in the light of what he has studied in the classroom, and thus the laboratory would become a place for empiricism instead of a place for thought.

Third. If the tests are not so arranged that they come after the class-room work upon this same subject, the students will have a very narrow and vague idea of the principles involved, and the laboratory work performed will have but little educational value.

The conduct of the laboratory should be planned with the same objects in view; and for this reason the following rules should be adhered to:

First. There should be a sufficient number of competent instructors in the laboratory, so that each squad of students making a test of any kind may be so thoroughly looked after that there shall be the certainty of good work. When the students first come into the laboratory, they are only capable of taking a few observations in a given time, but as they become more used to testing, more can be given to them to do, so that at the end of the year three men can do what it would require six to do at the beginning. In our own laboratories, during the third half year, the students, as far as possible are made to take some responsibility, *i. e.*, they start apparatus and make tests without help from the instructor who simply watches them and keeps quiet as long as everything goes right. While the students like this arrangement and profit by it, it can not be carried out with all the apparatus.

Second. Throughout all portions of the student's work in the laboratory, and throughout all the tests the greatest care and accuracy should be insisted upon, and no careless work whatever should be tolerated, the student having assigned to him only so much work as he can do thoroughly. A man who is trained to perform his work with accuracy can adapt himself to rough and ready work when the occasion demands it, but if he is not so trained, he can never be trusted to perform nice and accurate work, and hence he is not a safe man to be given responsibility.

Third. Each student should be required to make all the calculations for all the tests in which he takes part, and to hand in his results in a moderate length of time after the test is made; and not only should this work be examined by an instructor, but the instructor should work up the entire test himself independently, before examining the work of the students.

In our own laboratories all observations are recorded on large logs which are delivered to the instructor at the end of the test; and he works them up and puts his results at the bottom. Before delivering the large log to the instructor, however, each student copies off on a small log all the data he needs to work up the test, which he works up and hands in on the small log. These small logs, after being corrected by the instructor, are returned to the student who thus gets in a convenient form a summary of all the tests made by him. The amount of work required of the instructors is, of course, very considerable; at times there have been as many as eighty tests per week to be worked up. By following the method of procedure outlined above the results of the tests are reliable for they are to all intents and purposes, as far as their reliability is concerned, tests made by competent instructors using students as assistants, and so arranging the work of these assistants that the work may be accurate throughout.

So important is it that there should be a close co-ordination between the laboratory, and the classroom work, that it is very desirable that the one in charge of the former should also carry on some of the latter, for, in this case he can better appreciate what a proper co-ordination means, not merely in the case of the class-room work which he carries on himself, but also in the case of that given by others. Moreover, if he is a competent man this arrangement is possible even when the number of students using the laboratory (as in our own case) is very large.

In the case of the engineering laboratories of the Massachusetts Institute of Technology there is one assistant professor (Prof. E. F. Miller) in charge of the laboratory; he also carries on some class-room work, but for the laboratory work alone he has five assistant instructors. He takes pains to co-ordinate all the laboratory work with the work done in the class-room in the various courses, so that the laboratory exercises on any one subject shall come after the work done in the class-room upon the same subject. Each squad appointed to any one test is carefully supervised by Prof. Miller or else by one of the assistant instructors. One result of the above stated co-ordination is that when a certain squad begin a new kind of test, five or ten minutes suffices to give them all the information they need as to the arrangement of the apparatus, location of valves, etc. In the case of the more complicated apparatus drawings or sketches or blue prints of the apparatus showing sections, etc., are kept in the laboratory, accessible to the students at all times; and, as they have already studied the subjects involved, in the class-room, the preliminary explanations needed are few. Then every student in the squad is required to calculate and hand in the results of the entire test within a week. These calculations are examined by the instructor, and, if there are mistakes, the student must correct them and hand in the calculations again.

While there are provided a number of blanks for working up the results, these are such as would be required in practice, and only indicate the nature of the results required, and not the mode of making the calculations, the student being expected to know the latter from the principles which he has learned in the class-room. I ought to add that before the student begins working in the engineering laboratories at all he should have practice in physical manipulation in the physical laboratory, so that such things as the reading of verniers and micrometers, the reading and calibration of thermometers, however they are graduated, may be familiar to him, and that he may have acquired a certain freedom with manipulating physical apparatus. When the students first enter the laboratory the instructor should give them one preliminary talk explaining how the whole work is to be carried on, in other words, to acquaint them with the rules and customs of the engineering laboratories.

We may express all the proceedings briefly, as follows:

First. The laboratory work should not be conducted independently of the class-room work in the various departments, but should be closely co-ordinated with the latter.

Second. The student should be required to do thorough and reliable work throughout, and whatever amount of supervision is needed to attain this result should be provided.

Third. The student should be required to work up the entire set of results of every test in which he takes part, and to hand them in in a moderate length of time.

Fourth. An instructor should supervise the test as carefully as would be necessary if he himself were making it professionally as an engineer, and employing the students to assist him as observers.

Fifth. An instructor should work up the entire results of each test before the calculations of the students come in to be examined.

Sixth. The blanks furnished should contain only such headings as are necessary in a test of this kind, and none should be inserted for the sake of indicating to the student how to make the calculations, as he should know this from his class-room work.

With the regular laboratory exercises conducted in this way, the results will be reliable, and valuable, the student will have been drilled in thorough and accurate work, and the laboratory will aid materially in showing him the bearing which his class-room work has on practice, and he will have the best kind of training for any experimental work he may subsequently have to perform.

Besides this, the laboratory furnishes the means of making special investigations but of this I will not speak here but will merely refer to an address I made before this Society last year upon the subject of graduating theses.

11

EARLY INSTRUCTION IN PHYSICS AND MECHANICS.

By C. M. WOODWARD,

Dean of the Washington University School of Engineering,
St. Louis, Mo.

"Facts before reasoning" should be the teacher's motto. By facts I do not mean general laws which can be expressed by formulas; I mean such facts as the pupil readily sees and unhesitatingly accepts from personal observation and examination—the simple phenomena of matter. The study should be phenomenal, then rational; quality first, and quantity only when the methods of expressing quantity have been fairly mastered.

General formulas should be postponed till the students are familiar with the discussion of algebraic problems. The laws of falling bodies under ideal conditions are of little value. They furnish a fruitful field for "problems," which are of no earthly use, and which more or less mislead when the pupil comes to consider forces other than gravitation, and directions other than vertical.

The composition and resolution of forces should follow plane geometry, and the general propositions of statics should not be considered till the student is at home in Trigonometry and co-ordinate Geometry.

Attempts to anticipate mathematical methods by showing their use in practical experiments are confusing
and in the end unprofitable. The doctrine of static
moments is simple, and as all calculations are arithmet-
ical, experiments in that field can be introduced
early.

In my judgment the most serious blunders in
teaching mechanics comes from a premature introduction of the subjects of work, energy, and the effect of
varying forces and couples. These things are out of
place in elementary mechanics; they should be taken
up only after the student has mastered the fundamental
doctrines of the integral calculus. To be sure the student can calculate certain relations between times,
distances and velocities, from formulas established by
integration, but the work is not satisfactory; his mind
is not clear and the work does not appear to him
rational. The time is largely lost, and when he
approaches the subject from a higher plane, he enjoys
no advantage from the fact that some of the terms are
familiar, and some of the formulas have the appearance of old acquaintances. On the other hand he is
conscious of having been badly treated by being led
beyond his depth, and made to feel somewhat *non
compos mentis* because he could not keep his head above
water. Students often feel very indignant when they
come to fully realize the illogical methods they have
been made to follow.

I am a firm believer in the value of problems in
stimulating mental activity and leading to an apprecia-

tion of what has already been learned; but in early physics and mechanics, the problems should be solved in the laboratory. They should not be algebraic, nor geometric; they should be solved by investigation, by the use of old and new apparatus, by drawing additional conclusions from old experiments, and by new combinations. After a series of laboratory exercises, nothing can be more stimulating than a demand for the details of an experiment or a series of experiments which shall answer a certain question or establish an important fact. This study of design is educationally valuable in every way, but it must not be undertaken until the student has had a generous course in assigned experimental work. Design is an elaborative as well as a creative process, and the designer must have material to work on and work with.

I rejoice to see the rational methods of physical study now becoming popular. I regret only to see occasionally an ambitious attempt to omit the early and—to the teacher—very elementary steps and to spring at once to the consideration of quantitative relations which involve an experimental basis which the student lacks, and a mathematical analysis which he can not comprehend.

The object of these remarks will be attained, if writers of manuals, teachers, and directors of laboratories will listen to the words of an old teacher, and subject their methods to a rigid re-examination from the standpoint of the student.

TEXT BOOKS CONSIDERED AS SUCH AND NOT AS WORKS OF REFERENCE.

By C. H. BENJAMIN,

Professor of Mechanical Engineering, Case School of Applied Science,
Cleveland, Ohio.

I wish to present this subject briefly, rather as an introduction to a discussion than in the guise of a formal paper. The limits of my experience with text books advise me to confine this paper to the consideration of those which deal with engineering subjects. There is a difference of opinion among teachers as to the relative advantages of the text book and lecture systems, some preferring to do without books to the greatest extent possible. A perusal of the proceedings of the Chicago meeting of this Society, shows in the main an agreement on this subject which may be summarized as follows:

That a text book, or at least printed notes, should be used as an outline to the study of a subject and as a reliable record for the students' after use; that recitation work should be enlivened and vitalized by lectures and by informal talks together with such objective illustration as the subject may admit of; and finally that students' notes are a delusion and a snare. To convince myself of the last statement I have but to

look at the fragmentary records of my own college
days, which the added experience of later years scarce
enables me to decipher. Every term an examination
of the lecture notes of my engineering students reveals
such a meager record of data which have cost me much
trouble in preparation, as confirms me in the resolve to
put everything of the kind in print where possible.

The average student has neither the ability nor
the inclination to make a clear and correct report of a
spoken lecture. It may be further said of lectures as
a constant means of imparting information, that they
involve an unpardonable waste of time, requiring
double the amount to cover the same ground.

Take up a catalogue of works on engineering and
you will not find a subject unrepresented, while the
more common occur many times; but we all know to
our sorrow in how few of this multitude of counsellors
is there safety. Commonly the desire to make the book
salable among a larger class of customers has led the
author to introduce many data and statistics which are
entirely out of place in a text book. The very mass
of material which is of such value to the trained
engineer whose experienced eye at once sorts out the
desired information, is a slough of despond to the
student who flounders around without being able to
fasten on a single idea. And then the rambling, con-
versational style which so many authors adopt so
obscures the central idea that it escapes the pupil
altogether.

Descriptive and historical matter is allowed too much space in most engineering text books. The object of school training has been well said to be "teaching the boy what he can not get outside," and the school years are all too short to be spent in looking at pictures of engines and machinery, and reading lengthy descriptions, when the same information can be more readily obtained and more firmly fixed in the memory by a visit to a shop or a round-house. The examples or problems which we frequently see in such books are so absurd, so far from any actual conditions that could arise, as to discredit the whole book in the mind of a young man who has any practical knowledge of engineering.

The common formula for compounding a text book seems in many cases to be: Ten per cent. principle, twenty per cent. government reports, twenty per cent. irrelevant remarks and about fifty per cent. trade catalogues to fill up the desired number of pages, and make a book which will retail at five dollars, when the real essence of the subject could be put in a pamphlet and put on the market for a dollar. The excessive cost of engineering text books is a serious matter to many students, and when they realize as most of them do sooner or later, that they are paying for a large amount of padding, they naturally feel dissatisfied.

A text book, pure and simple, is a book to be used in instruction of learners and should be prepared with that motive alone. It need contain nothing original, but a presentation of the subject to be taught in a clear, concise, and above all in a logical form. The teacher should be able to say to his class: "Young gentlemen, there is not a paragraph in the lesson but what should be carefully read and thoroughly understood." If, for any reason, extraneous or supplementary matter is permitted, it should be either put in separate chapters or in smaller type that it may be distinguished from the body of the book. The treatment of the subject may be exhaustive or elementary, according to the class of students for which the book is designed, but it should be logical, direct and to the point without any rambling desultory remarks.

The mathematical treatment of engineering problems should be simple and practical in its character, avoiding "mathematics for mathematics' sake," and leading directly to the desired result.

Many books on engineering subjects are so cumbered with mathematical analyses which are only interesting to the mathematician as to render them unfit for ordinary use. The exhaustive general treatment of an engineering problem is often entirely appropriate for a work of reference, when a simpler solution covering the special case considered would be much more valuable to the student. The examples should be common-sense and practical in character and should be so worded that their exact scope and meaning can not be mistaken.

There is a good demand in this country to-day for concise, pithy manuals on nearly every engineering subject, which can be sold at a reasonable price, and there are plenty of men eminently qualified to write them.

MINIMUM LABORATORY WORK AND EQUIPMENT IN A CIVIL ENGINEERING COURSE.

By DWIGHT PORTER,

Associate Professor of Hydraulic Engineering, Massachusetts Institute of Technology, Boston, Mass.

Laboratory work has so great value for the engineering student as to be practically indispensable in a good course of education. It increases the interest of the student in the subject illustrated, and the instruction of the class-room is more deeply graven upon his mind than it otherwise would be. He feels that in the laboratory he is applying theory, and not merely learning it, and that he is actually dealing with some of the forces of nature. His capacity is strengthened for grasping and properly weighing the commonly accepted theories which are presented to him, and the relation between theory and practice is better appreciated. The student gains familiarity with the construction and behavior of certain standard forms of apparatus with which he is likely afterward to come in contact. He further acquires knowledge of the general mode of procedure and of the precautions to be observed in experimental work, and his self-reliance is increased for undertaking such work in after life. Training is received in observing and recording physical phenomena, and appreciation is

169

gained of the limitations, errors, and eccentricities pertaining to such operations. The laboratory offers wide opportunity for prosecuting useful and often original investigations in new fields, in connection with thesis and postgraduate studies. I think it is right to say also that the interest of the instructor and the efficiency of his teaching are directly enhanced by his contact with the laboratory and the work therein. Being thus attractive and useful alike to student and to instructor, the laboratory is therefore helpful to the school, in enabling it properly to perform its work, and in drawing students to its doors.

As I now look at the matter, the principal advantage of laboratory work for the undergraduate student should lie in and be measured by its ability directly to supplement, and to increase the efficiency of the theoretical instruction of the class-room. This advantage seems to me to furnish the only necessary and sufficient principle to be followed in installing and developing a laboratory plant for undergraduate use. Even aside from this the training of the laboratory is not to be despised. It is an education in itself of eye, hand, and mind, and a certain facility in its methods is doubtless of pecuniary value to the graduate who may be called upon at once to assist in similar work. Nevertheless, the value pertaining to engineering laboratory work as a sort of manual training, or as an end in itself, seems distinctly inferior to that lying in its use as a direct aid to theoretical instruction. The latter advantage will be sufficiently accompanied by the former in any well arranged and fairly equipped laboratory course. If this be the proper view to take, the laboratory equipment and practice of any school should be governed by the nature and amount of instruction given in the class-room. This in turn may vary with the location of the school, with the preparation or the needs of its student clientage, with its financial resources, or with its arbitrarily determined policy. One school may naturally, and to advantage, emphasize structural work in its curriculum, another railroad work, another hydraulic work, and so on, and the laboratory equipment and work should be consistent with the class-room instruction. But the advantages to the student in inspiration and in education, and to the school in true success and in reputation are so great from properly conducted laboratory work, that to ignore or slight it is to reject an aid to the highest results and to assume a heavy burden in the competition with other schools. At the same time the fact should not be overlooked that fine laboratories do not alone make a truly successful school. In my judgment a reputation for laboratories overshadowing the class-room and the work therein would be an unfortunate distinction.

It is difficult to make precise statements as to how much apparatus or how much time properly belong to a minimum laboratory course. As regards the time, however, I would hazard the opinion that the laboratory work with its accompanying computations should seldom occupy one fourth the amount given to the related class-room exercises and preparation, and that

30 hours usually will be found sufficient for a distinct line of work such as in Strength of Materials or Hydraulics. If the aim of the laboratory for undergraduates is mainly instruction, as I believe it should be, rather than results of scientific value, then in any one line of experiment the work should be carried but little farther than is necessary to assure the instructor that the student well understands what he is engaged upon and that he is reasonably proficient in manipulation and in observing. The object should mainly be to illustrate important principles, and repetitions or variations which do not serve to introduce new principles should be avoided. When earnest students, desiring the highest results from their efforts begin to ask "why should we perform this experiment, which teaches us nothing new," it is well to see whether work is not being given which had better be omitted.

I intend to speak of laboratory work for the civil engineering student, and shall present memoranda of what seem to be practically essential pieces of apparatus, with some approximate statements of cost. The number of pieces might of course be enlarged to advantage from time to time, and generally would be enlarged if the nucleus were once formed. But in the departments considered I think a school can not properly claim to have more than the rudiments of laboratories until it has the apparatus mentioned or its equivalent. It is here assumed that the student who comes to the engineering laboratory has already received a preliminary training in the physical and chemical laboratories of the school, where he has learned habits of accurate observation and has become familiar with sundry instruments and methods which he is incidentally to employ. Though considering the work of civil engineering students, I desire not to lose sight of the fact that in many respects there are common needs for all the prominent subdivisions of engineering—mechanical, electrical, etc. The existence of a mechanical, or even of an electrical or biological laboratory in the school may make it worth while to give the civil engineering student certain lines of experimental work, his needs for which would not alone warrant the installment of special apparatus.

TESTING LABORATORY.

Acquaintance with the physical properties of materials of construction lies at the basis of all engineering work, and the laboratory for testing such materials is therefore to be mentioned as the first essential. Here the student subjects proper specimen pieces to pulling, crushing and bending, noticing their behavior through all stages of the operations, and making the appropriate computations. In this, as in other laboratory equipments, the apparatus should be of sufficiently large capacity to permit of tests on a scale having commercial value.

This laboratory should contain the following pieces:

(a) A machine for testing pieces under tension or compression, and having a capacity of from 50,000 to 100,000 pounds. Such a machine should have suitable compression platforms, if compression tests are to be made, and can probably be purchased for from $700 to $1,000, according to its capacity.

(b) A transverse testing machine, with a span of say 16 or 18 feet, and a capacity of 18,000 or 20,000 pounds. A machine suitable for the purpose may be made to order for $250.

(c) In connection with the above machines, an apparatus for measuring the deflection of beams, and also one for measuring elongation or compression. These pieces may be made to order for $25 and $100, respectively.

(d) A cement testing machine which may be had for $125, more or less.

HYDRAULIC LABORATORY.

In hydraulic engineering the necessity is constantly arising for the measurement of the volume of flowing water, for testing the accuracy of devices applied to that use, and for determining the efficiency of pumps, motors and other devices. It is important, therefore, that the hydraulic laboratory should contain apparatus of sufficient variety to illustrate the different standard methods for the measurement of flowing water, whether free or under pressure; one or two of the best types of motors; and at least one or two standard forms of pumps. In the use of these the student will learn the difficulties of precise measurement of head, pressure, velocity and quantity; will learn to appreciate the relations between pressure and velocity in closed channels; and will observe the irregularities displayed by flowing water.

I would therefore specify for a hydraulic laboratory:

(a) A standard weir, its length dependent upon the quantities of water to be dealt with. This, with hook-gauge, weir-box and screens, need not cost more than $50 or $100.

(b) A Venturi meter, with connections to gauges for determining the coefficient of discharge and the friction loss.

(c) One or more hose nozzles arranged so that the pressure can be read and the apparatus used as a meter. These nozzles, with piezometer connections, will cost $10 or $15 each.

(d) One or two measuring tanks of at least 200 cubic feet capacity each, to be used for direct measurements of quantity and for calibrating other measuring devices. A good cylindrical steel tank of the above capacity can be made for $150 and a wooden tank for perhaps one third that price.

(e) A Pelton water-wheel of not less than two feet diameter, suitably arranged with gauge, brake, etc., for efficiency tests.

(f) A turbine, if possible, although the smallest commercial sizes of most turbines require a greater supply of water than it may be convenient to furnish in the laboratory.

(g) One or more pumps of common and approved form, with a total delivery of five hundred or more gallons per minute against at least fifty pounds pressure. These will serve not only to supply apparatus, where there is not a sufficient direct supply from without the laboratory, but also for efficiency tests.

(h) A set of floats, either tube or sub-surface or both, for use in outside measurements of discharge in flumes or streams. These will probably cost from $3 apiece downwards, according to style and size.

(i) An approved type of current meter for the same use, costing from $100 to $200, according to type and accessories.

(j) It is desirable also to have some form of Pitot tube, but perhaps this need not be considered an essential. A small Darcy-Ritter tube, for use in shallow channels, can be purchased for somewhat less than $40; and other forms, adapted to laboratory use and embodying the principle of the Pitot tube, can be made at a greater or less cost than the above, depending upon the details.

Piping, piezometer connections, gauges, chutes, etc., are required in fitting up a laboratory and may form an important item of expense, but their cost depends upon the particular arrangements adopted and will not here be estimated.

LABORATORY WORK IN CONNECTION WITH SURVEYING.

In the ordinary use of surveying instruments with classes of students we have example of a sort of field laboratory work so old, so general, and considered so necessary in civil engineering schools that I need do no more than refer to it. But I desire to call attention to the opportunity that exists for certain kinds of indoor work in connection with a course in surveying:—

(a)   In the use of the collimator for adjusting the line of collimation in transits and levels, and adjusting the standards in transits.   The apparatus required is simply two transits and a level, which are available in the outfit of almost any school, and fixtures which may be put up at a cost of $10 or $15.   The advantages are considered to be that adjustment of instruments, a more or less tedious operation, can be carried on at any convenient time indoors, and that the student grasps the principles involved better than in the usual out-of-door method, although he ought easily to make the adjustments by the latter mode after having worked with the collimator.

(b)   In the use of the level-trier, an arrangement for ascertaining the angular value of a division of the bubble tube, and costing about $40.

(c)   In the further practice that may be had in determining the magnifying power of telescopes, in inserting bubble tubes and in setting cross-hairs, no outlay of consequence being required for this work.

GEODETIC LABORATORY.

For a course in Geodesy, of practical value—one that should train the student for the work of the Coast Survey—it is believed that the following laboratory equipment is required:—

(a)   A direction instrument, for measuring horizontal angles, with micrometer microscopes in place of verniers.   A small instrument, not suited to actual field

12

use will serve in the laboratory, and can be bought from foreign makers for probably $300 or less. The student should be trained in the general use of the instrument and in examining it for all errors.

(b) A comparator, to be used in standardizing bars. From its use the student learns, as perhaps in no other way, to appreciate what a standard bar is.

(c) A small observatory, with portable transit instrument, chronometer, and chronograph. This outfit would be used in making observations for time, in connection with pendulum observations, etc., and would cost approximately $1,000.

(d) A small, half-second pendulum such as is used by the Coast Survey to determine the differential value of the force of gravity. It is believed to be best not to attempt with students to find the absolute value of g. A Kater's pendulum may, however, be used to advantage for the purpose of illustration merely.

(e) A magnetometer for determining the intensity of the magnetic current, and a dip circle for finding the value of the vertical component of the earth's magnetic force.

OTHER LABORATORY WORK.

In steam engineering the instruction usually given to civil engineering students is scarcely sufficient alone to warrant the installment of an experimental engine. If a school has a steam laboratory in connection with a course of mechanical engineering, then instruction to civil engineering students in valve gears and indicator cards may be supplemented by laboratory practice in setting valves and taking cards, for which purposes an engine of 10 or 15 horse-power is suitable. For complete engine testing, however, a course of instruction in Thermodynamics is required, and a large experimental engine of advanced type is desirable.

The equipment that has been described in this paper is believed to be necessary to accompany properly the instruction of the civil engineering student in the branches named. It comprises apparatus that is likely to be permanently useful, and should form a suitable basis for future development of the laboratory, whether for undergraduate or postgraduate needs. If a school have special laboratories other than here considered, then advantage may perhaps also be taken of those, but in any case let the laboratory work be proportioned to the instruction of the class-room.

A FEW MISTAKES IN THE CONDUCT OF COLLEGE FIELD PRACTICE.

By O. V. P. STOUT,

Adjunct Professor of Civil Engineering, University of Nebraska, Lincoln, Nebraska.

There seems to be reason to believe that all will agree essentially with the statement that one result to the student of a course in civil engineering should be that he will be able to say justly and truly, to himself, after graduation, and on the occasion of his first day's work in the field. "I know the principles which govern in the performance of this work, and am perfectly familiar with the construction and adjustment of these instruments; I have even a considerable knowledge of the details of such work, and have acquired some degree of skill in handling the tools and instruments employed, but I know that in a college course there is no opportunity to present all details nor to familiarize students with all of the various conventions and customs which are practiced in the best work in different parts of the country. I have been trained to observe, and to keep well in hand whatever work I undertake; hence, I believe that even though it is an advantage to me to occupy now one of the lowest positions in the party and thus have opportunity to gain by observa-

180

tion of the work of others some knowledge which I now lack, that if placed in a higher position, as for instance that of instrumentman, that I could discharge at least the routine duties of that position with reasonable satisfaction to my employers and credit to myself."

That all new graduates, even those of ability, do not approach their first work in this spirit goes without saying. On one side of this happy mean of firm and justifiable self-confidence, and realization of one's own powers and limitations, we find the extreme of great conceit, and on the other the opposite extreme of almost complete lack of self-confidence. As to the first of these, it is a pleasure to be able to say that it is so rare (some very good stories to the contrary notwithstanding) and the growth of it is so easily avoided in all students except those having a fixed tendency in that direction, that no special treatment is required here. As to the second, an instance comes to mind of a young man from a state university who actually implored a chief of party to permit him to take a chain or a rod instead of the transit which it was found necessary to put in his hands on his first day in the field. Aside from a few mistakes due to nervousness, his work on that day showed his lack of confidence to be wholly without cause. Here, it seems, was a case wherein a young man of more than average ability had been given excellent instruction and then had been permitted to doubt his grasp upon the knowledge acquired, and its direct applicability to practical work. Such a frame of mind can be accounted for in various ways. One man who was thus afflicted in his first experience attributes it to his having heard and believed too much of the harping on the extreme divergence of theory and practice, so that he was quite unprepared to find that work was done in somewhat the same manner as that described in the books and taught in the college. There are many such cases, and unless very careful an instructor may inadvertently lend encouragement to such ideas.

One practice that leads to unsatisfactory results in college field work is that of assigning work to students with the object in view of leading them to some particular blunder. This is done with the idea that the student will act here as he will later in practice, and see to it that he does not make the same mistake twice, but it is found that the matter can be overdone, and that after falling into a number of such traps, many students become, according to temperament, discouraged, uncertain, or careless and indifferent in their work, and at a subsequent opportunity may repeat the same blunder. It can not be assumed that such incidents make as permament an impression in the college field work as in the outside work where a more serious consequence is liable to be visited upon the one responsible for a single error. As pointed out at the Chicago meeting, a party of students working alone may make errors which destroy the value of their work, or may reach a point where unable to proceed at all, when a mere hint from the instructor, if present, would set them right. It is a mistake to take the time of the party or permit it to be taken to any great extent, to emphasize the error or disadvantages of some particular method or lack of method in doing work. The fact is that time is saved and better results are secured by performing an operation twice as it should be done, trying to make the results of the second operation better than those of the first, and securing a *drill* in the right way of doing. If permitted, for a considerable length of time, to proceed in a wrong manner, the student is more than likely to fail, even upon suggestion, to review the whole transaction in his mind and note the point at which he departed from the right path; so that, on attempting to repeat the operation he may go wrong at the same point as at first. It must be borne in mind on the other hand, however, that a too immediate and minute direction of work by the instructor limits the opportunity for full development of the student's powers and puts him in a wrong attitude toward his work. Enough responsibility should be placed on him to make successful prosecution and completion of the task assigned a ground for the justifiable self-confidence, which it is desirable should be implanted. Enough of the probable wrong turns, and pits, and snares should be pointed out to leave so few as not to distract at first, nor lead later to indifference.

I believe that there is imperative need in a civil engineering course of a drill in field work, extended considerably beyond that required merely for exemplification and illustration of the principles taught in the class-room. Even in these days of scanty demand for engineers it frequently happens that a man but recently graduated is placed at once where he is expected to handle instruments, and to direct a certain amount and grade of work skillfully, even mechanically. A young engineering graduate of a prominent university worked for a few months as assistant topographer on railroad location. Here he acquired a good knowledge of align- ment, but had no practice in the use of instruments. Then he was made transitman in another party, and handled his instrument so awkwardly during the first few days that it took several weeks of excellent work on his part to raise him to the proper level in the esteem of the members of the party. The idea expressed in the first sentence of this paragraph must have occurred to him with force.

A field engineer, on commencing the work of instruction, is likely to attempt to reproduce too liter- ally in that work the actual conditions of outside field practice. From experience I know that this is a mis- take. Many of the conditions and surroundings of actual field-work are distracting to those who have not had the drill, day after day, that enables them to do a large part of their work almost mechanically. Phys- ical fatigue should be avoided as much as possible. A young man who is tired physically is not at the same time very active mentally. Students should be encour- aged to do work as expeditiously as practicable; but too much hurry, while it may accomplish the staking out, or whatever the work is, deprives the student of the opportunity for full appreciation of the significance of the different steps of the process.

If the number of students in a field party is kept down to the small limit recommended by some, it seems that a certain discipline and practice in directing men and work is lost. A student, placed in charge of a large party, occasionally shows traits which were unsuspected while he was in charge of a small party. A student whose work with a small party was nearly always better and more quickly done than that of his fellow, found the case reversed when they were put in charge of parties comprising eight or ten men, with the organization of the party and direction of the work to attend to.

In concluding this paper, one reason suggests itself why the field practice of an engineering course should be conducted with infinite pains, to the end that the graduate will be equipped as indicated by his supposed statement in the opening paragraph. His first and perhaps his most lasting impression of the worth of his college training comes to him quite generally from the degree of success which he encounters in his first work after graduation. As this, in a majority of instances, is some class of surveying, it is only just to the student himself, to the instructor, and to the institution, that this training in this subject be such that he can not look back after a few years of practical experience, and point with justice to mistakes in the conduct of his college field practice.

THE NEED OF MORE EXTENDED PROFESSIONAL STUDY, WITH SPECIAL REFERENCE TO PRESENT COURSES IN STRUCTURAL ENGINEERING.

By EDGAR MARBURG,

Professor of Civil Engineering, University of Pennsylvania, Philadelphia, Pennsylvania.

With the rapid advancement of civil engineering practice, the arrangement of a course of general preparatory instruction is becoming a matter of increasing difficulty and responsibility. The ideal course of professional study is unfortunately dependent, for its actual institution, on conditions which are themselves largely ideal and which have only been attained to a confessedly imperfect degree in the best of our technical schools. There are manifestly three prerequisites to the realization of ideal educational results—the educators, the students and the material equipment of an institution must all be of the ideal pattern.

The engineering educator should be primarily a man of broad culture, with profound theoretical attainments in what may be termed the science of engineering. He should have acquired, moreover, a certain familiarity with the practical details of engineering operations in general, combined with a most intimate acquaintance—that kind of acquaintance which can result only from extended experience in actual practice —with one of its specialties. His educational sphere should then be limited strictly to the treatment of subjects directly related to his own specialty and he should be afforded ample opportunities for continued professional practice. The interests of the institution with which he is connected and his own would be mutually best served by such an arrangement.

A department of civil engineering modeled on these lines, would require the services of at least four, better six or eight specialists, with a proper complement of assistants, to render its technical work thoroughly effective. There can be no doubt but that, with ample financial support, an organization of this general character might be readily perfected. The question of material equipment resolves itself even more directly into one of an essentially pecuniary character. There are, however, few if any professional schools on a sufficiently independent financial basis to fully inaugurate such a policy. Certainly no institution in this country can justly be credited with a staff of educators of the indicated standard in the entire field of civil engineering practice. These reflections, serious as they may appear, lose much of their significance when considered in due correlation with the difficulties surrounding the attainment of what has been cited as the third prerequisite to the realization of ideal results, namely, an ideal grade of preparation on the part of the prospective student. The ultimate adjustment of this matter—whether through the rigid exaction of greatly increased requirements at entrance, the extension (not specialization) of the professional course, or by the gradual improvement and better systematization of the preparatory schools—can be, as yet, little more than conjectured. Certain it is, however, that under conditions now prevailing, the time that can wisely be devoted, in a four-year course, to instruction in the actual methods of more advanced engineering practice, is exceedingly limited.

In every engineering course, regardless of its extent, the first effort should undoubtedly be directed towards laying, conscientiously and well, that broad rational foundation to the future engineer's education, on which alone the delicate and complex superstructure can be safely and logically reared. Principles first, principles second, methods third, should be emphatically the motto of the modern engineering school. On the other hand, the writer is unable to concur with those who hold that the province of technical education can not, for the best interests of the student, be extended much beyond the field of pure theory, with such limited and imperfect practice in its application as is now afforded in the more advanced institutions in this country. It would seem, on the contrary, that, in this age of continually narrowing specialization, there is an ever increasing need of broader academic training in matters of practice, provided that *such training be given under the immediate direction of a corps of competent specialists.* If such opportunities were offered—and the writer has an abiding faith that this time will come—may it not reasonably be expected that the graduate will be more than compensated for the expenditure of an additional year or two in preparation? Is it not a matter of common observation that there is a growing tendency on the part of young practitioners to settle down permanently in certain specialties without experience in other professional lines and subjected to all the disadvantages accruing inevitably from such deficiency? Under existing conditions, this collateral experience can, in fact, only be obtained in so loose and haphazard a fashion, that the time spent in its acquirement is often, if not usually, out of all proportion to the real benefit derived. If this be conceded, should it then not be accounted as one of the highest functions of the truly professional school to avert so far as practicable the danger of such ill-balanced development, by modeling its technical courses on the best and broadest lines?

In a recent series of articles, edited by "Engineering News," it appeared from the average returns of fifteen leading schools of civil engineering, that about 22 per cent. of the total time of instruction was devoted to advanced technical work, including under this heading the following nine subjects: Thesis, Materials of Engineering, Mechanical Tests, Steam Engines and Boilers, Framed Structures, Masonry, Railways, Hydraulics and Sewerage. Considering only the latter five, as strictly civil engineering specialties, and excluding thesis work, as an indeterminable factor, the proportion sinks to only 12 per cent. of the entire course. If it be assumed, at a conservative estimate, that two-thirds of

this remaining time is assignable to text-book studies
and lectures, their remains only 4 per cent. to be credited to work in actual designing and other matters pertaining to engineering practice, of a character more
advanced than elementary drawing and surveying.
That this proportion is lamentably small, the most
ardent advocate of our present educational system will
probably admit. Its judicious apportionment may well
receive the most earnest and critical consideration.

The policy in vogue at our best technical schools
is doubtless the most advantageous one under present
circumstances; namely, to concentrate the time available for instruction in the methods of practice, principally upon two or three of the most important specialties,
supplementing this as far as practicable by short, well
arranged courses of lectures and conferences in the
treatment of other subjects. The writer has attempted
to present some statistics in this connection, by collat-
ing such data as could be gleaned from the various college catalogues. The character of the work and the
division of time was in many cases so imperfectly
defined, however, that a reliable synopsis could not be
prepared. Nevertheless, a few general observations
may be of interest.

From an examination of twenty-six catalogues,
including those of all civil engineering schools of prominent rank, it appears that a course in the designing of
hydraulic and sanitary works is offered in only two
instances, excepting a short course among the "options"
at a third institution. The design of masonry struct-

ures is found to enter the curriculum of only three of
these schools and, in a single case, provision is made
for a course in the designing of railway standards.
Twenty of these twenty-six institutions supply courses
both in the designing of framed structures and in railway office work, and, of the remaining six, all but two
offer instruction in one or the other of these two specialties. At a very rough average, based on these incomplete data, it seems that about three times as many
hours are devoted to structural design as to railway
work, which, for limited courses in both, the writer
regards as about the proper proportion.

The notable prominence given to courses in structural engineering appears to be well warranted, for the
following principal reasons:—

1. The civil engineer is pre-eminently a construct-
or. It is the correct application of the principles of
statics with which he is chiefly concerned. A sound,
scientific knowledge of the art of construction is essential to high success in any branch of his profession.

2. The special field of structural engineering is
itself on the eve of an era of most promising development. The five-hundred foot truss is but an achieve-
ment of yesterday. The seventeen hundred foot truss,
the twenty-two storied steel skeleton building and the
thousand foot tower are among the triumphs of to-day.
With the diminished cost of steel and iron and their
scientific adaptation to architectural requirements, stone
and timber are being rapidly supplanted in building

operations, though this movement is probably yet in its infancy.

3. Structural engineering affords an opening that is especially attractive to the educated engineer. In perhaps no other branch of the profession is it so difficult for one possessed of limited educational advantages to attain eminence. It is well-nigh impossible "to work up from the ax." Other things equal, the well-trained graduate has effectually outstripped his less favored competitor from the start. The technically self-educated seldom rise above the level of the draughting board.

4. With the single exception of railway engineering, there is probably no branch of the profession which enlists so large a proportion of civil engineering graduates. Complete statistics of their occupation during the first five or ten years of practice would be extremely interesting in this connection and if carefully continued for succeeding classes would serve to indicate fairly well the trend of the times. Very few schools, however, publish such records.

In the following table, the present occupations of the alumni of the Rensselaer Polytechnic Institute are classified for two five-year periods, from 1883 to 1892 inclusive. This table covers a list of 292 graduates. While the figures evidently do not warrant even approximately general conclusions, they are interesting and instructive, to some extent at least, from the fact they represent accurately the returns from the oldest and one of the most successful engineering schools in this country—an institution which offers to-day but a single, unspecialized course in civil engineering and which draws its students from so wide a territory, that the results are presumably not unduly affected by local or environing conditions. In a second table a more detailed sub-classification is presented of those engaged in strictly civil engineering practice, constituting about 45 per cent. of the entire number.

TABLE I:

OCCUPATIONS.	1883 to 1887 inclu.		1888 to 1892 inclu.		Total.	
	No.	Per ct.	No.	Per ct.	No.	Per ct.
Railway Engineers..................	15	9.4	12	9.0	27	9.2
Structural "	25	15.7	29	21.8	54	18.5
Municipal "	17	10.7	14	10.5	31	10.6
U. S., State and Canal engineers.....	1	0.6	6	4.5	7	2.4
In miscellaneous practice......... ..	11	6.9	2	1.5	13	4.5
Total in Civil Engineering practice..	69	43.3	63	47.3	132	45.2
Architects.........	3		2		5	
Mechanical engineers..............	0		0		0	
Electrical "	6		3		9	
Mining "	4		1		5	
Total in all lines of engineering practice and architecture...	82	51.6	69	51.9	151	51.7
Teachers of Science and Engineering	8		6		14	
Contractors.......................	4		3		7	
With manufacturing establishments	27		11		38	
Chemists and Metallurgists........ ..	2		5		7	
In other pursuits related to engineering..............	6		4		10	
Total in pursuits related to engineering...................	47	29.5	29	21.8	76	26.0
In business......................	10		14		24	
In other foreign pursuits	3		5		8	
Occupation not stated or not closely specified (engaged partly in engineering)..................	17		16		33	
Total in non-engineering pursuits, or not classified.............	30	18.9	35	26.3	65	22.3

TABLE II.

OCCUPATIONS.	No.	Per ct.
RAILWAY ENGINEERS :—	27	20.5
Consulting bridge engineers	3	
With bridge works...............	22	
" " engineers............	2	
" Elevated R. R..............	1	
Bridge engineer R. R. Co........	1	
STRUCTURAL ENGINEERS. Total on bridge work............	29	22.0
With architectural iron works....	6	
" " engineers......	5	
Total on architectural iron works.	11	8.3
Inspectors bridge and archt. work	10	7.6
Ship building...................	2	1.5
Masonry and foundations........	2	1.5
Grand total Structural Engineering............	54	40.9
City engineers..	7	
Assistant city engineers..........	2	
MUNICIPAL ENGINEERS. Hydraulic and sanitary engineers.	5	
Assistant waterworks engineers...	5	
" sanitary " ...	3	
Engineers street railways...	5	
" pavement construction..	4	
Grand total Municipal Engineering............	31	23.5
U. S., STATE AND CANAL ENGINEERS.....................	7	5.3
IN MISCELLANEOUS PRACTICE..........................	13	9.8
Grand total.............................	132	100.0

It is seen from these tables, that of those in actual
practice nearly 41 per cent. are employed in structural
engineering, without including those engaged in general architecture. This percentage is exactly twice as
great as that for railway engineering and about three-

fourths in excess of the aggregate percentage for all branches of municipal engineering. It seems probable however, as before intimated, that if the returns from all our technical schools could be thus classified, railway engineering would show the largest following, structural, hydraulic and sanitary engineering succeeding in the order named. It is worthy of at least passing note, that of the eleven engaged in architectural iron work, nine belong to the latter five year period, as do also the two employed in ship building. Marine engineering, or in a narrower sense marine architecture, deserves to be mentioned as another field, apparently destined to a great future in this country, for which a thorough knowledge of structural designing is of first importance.

The writer proposes, by your leave, to submit, in some detail, his views as to the conduct of courses in the theory and practice of structural engineering, the same being essentially a description of the methods pursued at the University of Pennsylvania during the past two years.

A course in "Mechanics of Materials" forms a logical introduction to the study of the more advanced theory of stresses. This course should serve primarily to thoroughly familiarize the student with the common theory of flexure and its practical application to the designing of beams and columns. Special care should be taken to bring the student to a clear appreciation of the fundamental basis and the consequent limitations of this important theory. To this end, his attention may

profitably be directed to some of the numerous erroneous deductions and loose statements to which it is continually giving issue, as witnessed in the technical press.
The writer lays great stress on the liberal resort to carefully selected problems.   Instead of depending mainly
on those given in the text-books, they may better be
varied from year to year.   The work can thus be more
suitably adjusted to the apparent needs of each particular class and incidentally the temptation of relying
unduly on the heritage from their predecessors, which
some students find so difficult to resist, is effectually
removed.   The more difficult problems may be assigned
as exercises to be solved at home and returned in neatly
written form.   Simpler ones should be frequently employed as impromptu work in class-room.   They afford
an excellent, albeit an ofttimes disappointing test of
the student's actual attainments.   It may be accounted
at once, as the highest duty of the educator and the
most difficult one, to accustom the student to habits of
close, self-reliant, logical reasoning.   The average student, after an apparent mastery of fundamental principles, will often evince a most unexpected and discour-
aging inaptness in their applications, when the typical
conditions are in the slightest degree modified or ob-
scured.   This difficulty can certainly be met in no other way so well as by giving him continued and varied
practice in the solution of problems.   The latter should
be given a distinctly practical trend throughout.   The
student should receive an extensive drill in the computation of shears and bending moments for simple and

continuous girders and in the proportioning of solid and flanged beams. Practice in the determination of centers of gravity, moments of inertia, etc., should be confined more especially to standard forms of built sections. The student's interest may be greatly stimulated, even at this early date, by bringing his work in the closest possible relation with the world of practice. His mind should be effectually relieved of any lurking suspicion that such matters as moments of inertia and radii of gyration are little more than text-book phantasies. Instead of merely emphasizing this in words, it should be the more forcibly impressed by exhibiting in class-room blue-prints and hand books containing extensive tabulations of these and other functions used in every-day practice. In due time, every student should be required to provide himself with a copy of the Carnegie Pocket Companion and initiated in the skillful use of its tables. The student will apply himself with renewed interest and vigor, when he finds to what an amazing degree he is relieved of what soon becomes little more than irksome drudgery. The time thus gained may be devoted with greater profit to the general analysis of more difficult problems. This policy may be pursued to good advantage throughout the course. On the completion of the theory in any subject, the conclusions should be duly summarized and presented to the student in a concise, practical form. After the study of the theory of columns, for example, he should be shown the diagrams in the Transactions of the American Society of Civil Engineers, and it should be clearly pointed out that, from the usual length-ratio at which a column begins to fail through flexure as a whole, to the upper limit permitted in good practice, the extremely simple, though strictly empiric, straight line formula yields results substantially identtical with those derived from formulas founded on a rational basis; that the agreement is, in fact, closer than necessary in view of the erratic variations that obtain inevitably between the physical column and its ideal prototype. It should be remembered always that we are engaged in the education of future engineers, not scientists. The accomplished engineer should doubtless understand clearly the derivation of the formulas he uses, in order to judge intelligently of their limitations. At the same time, to the busy practitioner, it is a matter of chief moment to adopt the simplest formula which yields trustworthy results within the limits of practice, regardless of its applicability at such extreme limits as may be set in the scientific laboratory.

On completing the study of "Mechanics of Materials," the student is qualified to enter simultaneously, say during the second term of the Junior year, upon courses in the analytic and graphic treatment of stresses in simple trusses and in the designing of plate girders, The analysis of continuous and partially continuous trusses, the theory of suspension bridges and elastic arches, as well as a more advanced course in structural designing may be left to the Senior year.

The writer thinks that Graphic Statics can be best presented as an entirely separate course, but that care should be taken not to give it undue prominence. Its application should be limited mainly to those types of structures for which graphic methods afford distinct advantages. Most structural engineers in this country will agree that in determining the stresses in all the ordinary forms of bridge trusses, both from static and moving loads, but especially the latter, analytic processes are, on the whole, vastly superior. But a stronger reason may be urged for subordinating graphic to analytic methods during the scholastic period. That reason is, in fact, implied in the very significance of the qualifying terms. The former is largely mechanical. It does not employ the higher faculties to nearly the same degree as does the latter. After a very limited amount of practice, the student will seldom be at a loss to do graphically what he can solve analytically. The reverse is by no means true, however. Graphic processes may be applied though the underlying principle be forgotten. With analytic methods, it is essential to know not only the *how*, but the *why*. A course in graphic statics is admirably well adapted to the cultivation of habits of accurate mechanical execution. The student finds himself obliged to gauge the accuracy of his own work by inflexible standards, moreover by standards which can usually not be applied until the work has been completed. If careless or unskilled, he stands self-condemned and must incur the self-inflicted penalty of repetition. The system is a truly ideal one. Unfortunately it does not admit of universal application. The writer does not wish to be understood as underestimat-

ing the educational value of an elaborate course in graph-
ics, as offered in some of the European schools where
the question of time is a less pressing one than with us.
He is clearly of the opinion, however, that under conditions here existing, an advanced course in this subject
can only be given at the expense of other subjects of
greater or more immediate importance.

The study of structural design may be advantageously begun with a course in the designing of plate
girders. This course may be conducted concurrently
with that in the theory of stresses. The writer regards
such a course as an essential part of the curriculum.
He thinks it should not be omitted, even though more
advanced work may have to be sacrificed. In this the
fundamental principle is recognized that courses of
instruction in the methods of practice should be
arranged primarily with a a view to fitting the future
engineer for such duties as are most likely to fall to his
lot during the period of his earliest practice. Judging
from results, this principle, hackneyed though it be, is
but indifferently enforced at many of our technical
schools. How is it otherwise to be explained that so
large a proportion of our graduates are wholly unskilled
in free-hand lettering, that so many lack speed and
accuracy in ordinary computations and approach the
simplest duties in an unsystematic, unworkmanlike
manner. These strictures are surely not exaggerated.
The writer has encountered men, holding technical
degrees from reputable institutions, who were clearly
unpracticed in the use of the duodecimal scale, and

many who were wholly unfamiliar with the use of the slide-rule. Granted that such matters may be readily picked up in practice, is it to be denied that the graduate is seriously and needlessly hampered at the outset of his career by such deficiencies? Where the evil and its corrective are both so palpable, may it not well be a source of wonder to practical men that these conditions continue to prevail to the extent they do? One hour a week devoted to free-hand lettering throughout the Freshman year, supplemented by practice incidental to the courses in map drawing and structural design, is sufficient to insure to the average student the requisite proficiency in ordinary lettering. The course in elementary drawing should consist largely of the construction of scale drawings, based on free-hand sketches. Accuracy, neatness and lastly speed should be duly insisted on.

The student should thus have developed into a fair draughtsman, before the course in plate girder designing is reached. The equivalent of two hours per week, during half the academic year, devoted to this course, is sufficient for the preparation of estimates for a deck and through design and a turn-table, together with general drawings for the first or second. The early part of the course may be devoted to lectures and during this period, definite portions of the computations for the design in hand may be assigned as exercises, to be subsequently discussed in class and finally entered in a note-book in a neat, systematic form. Every student should be required to provide himself with a set of standard specifications, not only for the sake of better training, but also to invest the work with a closer semblance to actual practice. Cooper's specifications commend themselves, by reason of their clearness, conciseness and general excellence and from the fact that they are most extensively used in practice.

The writer does not think it advisable to undertake the preparation of complete shop-cards. To do this thoroughly, consumes much additional time that might be turned to better account. Unless such work is precisely what it purports to be, it had better not be attempted. While the general dimensions of all parts should be indicated on the drawings and the necessary calculations should be made for the smallest details, such matters as the exact spacing of rivets in lateral connections, for example, may well be omitted. If the student were allowed free access to blue prints of a duplicate design, his own drawing might be completely elaborated, but the writer is decidedly opposed to that kind of instruction, which he regards as little better than more advanced exercises in copying. Towards the close of the course, and not until then, numerous blue-prints of selected examples from practice should be laid before the class for critical discussion. The good and bad features of design may be thus emphasized. It serves, moreover, as a wholesome revelation to the student to note the many different ways of accomplishing essentially the same results, in even so elementary a matter as the design of a plate girder bridge.

The course in more advanced designing may be conducted in the same general manner as that outlined.

At the University of Pennsylvania, four hours per week, in two periods, throughout the senior year, are allotted to this work.   This time is devoted mainly to the design in complete detail, of a through, pin-connected, Pratt truss railway bridge.   This is supplemented so far as the time permits, by lectures on the design of other typical structures and exercises in the computation of stresses.   In this course also, no models or reference plans of any description are placed before the student, until the work is well advanced.   Each detail is separately developed before the class in free-hand blackboard sketches, which are minutely discussed and serve the student as a means of general guidance in his work.   It is perhaps needless to add that he is required to keep a full set of notes which are finally submitted for examination.   No practice is given in the preparation of shop-bills and mill-orders.   The time is considered too valuable to be spent on matters of pure office routine, the details of which are not only extremely simple, but differ considerably at the various manufacturing establishments.   Such a course affords admirable opportunities for familiarizing the student with the many time and labor-saving devices used in practice, such as shear and moment diagrams of engines, tables of post and chord sections, of pins, rivets, etc.   Each student should be thoroughly drilled in the use of the Thacher calculating instrument, that invaluable aid to the busy computer. At the same time, he should be encouraged to habituate himself to performing minor operations mentally.   The importance of teaching the student to discriminate

intelligently between a proper and consistent degree of
accuracy on the one hand and unwarranted refinement
on the other, deserves to be especially emphasized. The
treatment of concentrated load systems presents opportunities in this connection which should not be left
unimproved.   Great stress should also be laid on the
value of approximate checks for the detection of important errors.

The writer can not refrain from expressing his distinct disapproval of certain methods in vogue, by which
the student is expected to take complete measurements
of some structure of his own selection and afterwards
to prepare detailed drawings based on sketches collected
in the field.   He has had opportunities of judging the
results from this form of instruction, and although these
are doubtless affected to some extent by attendant circumstances, it is his deliberate judgment that the
scheme is unsound in principle and defective in its
practical operation.*

A limited number of carefully planned visits of
inspection to structures complete and in process
of erection, as well as to manufacturing plants
should be regarded as essential auxiliaries to the
well-ordered course.   They tend at once to enhance the
student's interest in his work and to accustom him to
intelligent, critical observation.   Incidentally he should
be cautioned to be circumspect and deliberate in forming
his conclusions and warned against that reprehensible
habit of hasty condemnation and petty fault-finding,

---

*A number of the professors present expressed a disagreement with the author in
this matter.—ED.

so often characteristic of immature judgment. It is desirable that these tours be extended especially to such structures as, for example, swing bridges, which can be at best but inadequately treated within the limits of the present curriculum. The value of this kind of instruction may be greatly increased by procuring blue-prints of the original working drawings for careful review before the visits of inspection are undertaken.

There are a number of subjects, connected with structural engineering, which, partly through want of time and in part from their very nature are not susceptive of satisfactory treatment in the academic course. Their importance is such, however, that they should receive, at least, some general attention. To this end, near the close of the regular course, a series of special lectures may be presented, covering the fundamental elements of the following topics:

First. Principles of Economic Designing.

Second. Methods and Cost of Manufacture.

Third. " " " " Erection.

Fourth. Duties of the Inspector.

The subject of general designing should be approached from its æsthetic, as well as its economic side. It is a serious reproach to our profession that so little has been accomplished in this direction. Intensely utilitarian motives are the common characteristics of American designs. Even our city bridges, with few exceptions, are conspicuous monuments of ugliness. Though this is attributable partly to municipal parsimony, the burden of blame should doubtless rest with the engineer. It is presumably largely the fault of his early training. The artistic side of his education is usually wholly neglected and in practice there are few opportunities for subsequent development. A short, well-planned course in architecture, in competent hands, suggests itself as a partial remedy.

The writer has purposely refrained from any special reference to the teaching of specifications and laboratory work in mechanical testing. Both of these topics have been made the subjects of special papers at this meeting.

AMOUNT AND KIND OF SHOP WORK REQUIRED IN A MECHANICAL ENGINEERING COURSE.[*]

By C. W. MARX.

Professor of Mechanical Engineering, University of Missouri, Columbia, Missouri.

One of the most unsettled and debated questions which arise in framing a course of studies for an engineering curriculum, is the amount and kind of practical work, i. e., laboratory, field, and shop work. Probably the great difference of opinion which now prevails on this subject is due in large measure to differences in the training and in the personal experiences

------
[*]Read by title only for want of time.

of individuals after leaving the schools. Laboratory and field work are of a much longer standing custom and practice than is shop work, so that their importance in a curriculum is more easily settled, but when shop work is touched upon the disagreement is strong.

Shop work should occupy a coordinate and not a subordinate position in the mechanical engineering course; it should be a required and not an elective branch. The object of shop work is not so much to teach the students manual dexterity, as to train their hands and eyes while they are pursuing the other engineering branches. This mental training acquired by means of the hand and eye will enable the students to see matter and life from more than one side; they acquire the habit of working in a systematic, thoughtful, and thorough manner; they learn to think how work can be done expeditiously and economically, and then to work in accordance with their thoughts; they come to have a definite object in view while taking each step; in short, tool instruction and practice so trains the hand and mind that the one intelligently follows the dictates of the other. The object is, therefore, not to make a skilled mechanic of the student, but to give him the knowledge required by a foreman and superintendent, so that he may be enabled to direct the labor of others intelligently. Experience has repeatedly shown that a young man with such a training in addition to his engineering education readily outstrips his untrained competitors in the mechanical line. His aim and motto should be—not, "how can I myself do the work best?" but, "how can I direct the labor of other men and the machines to do it in the best and cheapest way possible?" His duty should not be that of a mechanic, but a director; not a follower, but a leader; not a soldier, but a commander. This however should not be the goal of a student's ambition, but should be only one of the intermediate steps to the position of engineer and designer; for it is the best channel through which to advance in order to become a safe, economical, and successful designing engineer.

There seems to be a general consensus of opinion among the advocates of shop work that carpentry and joinery, blacksmithing, and machine work should constitute the body of the course in shop work. In wood work, wood turning and pattern making should be included and this brings with it the art of molding and founding. Among the first things to be taught and impressed upon the student's mind in all the shops, is shop discipline, order, promptness, and neatness. The student's knowledge of the care, use, and abuse of tools progresses as do the exercises.

In the wood shop, at the bench, a series of fifteen carefully graded exercises should be made which gives the student practice in the use of the different tools and a knowledge of the ordinary joints used in the arts. Drawings of the various exercises should be made on the board or on large sheets of paper; these drawings should first be explained, then the instructor should perform all the new operations which involve new principles in the presence of the whole class. Each student should then make a drawing to a reduced scale in a blank book making all the necessary projections and even an isometric in special cases; special attention should be directed to the position of the necessary dimensions, no other instruments than a two foot rule, a square, and a thirty-degree triangle made in the shop should be used. After the series of exercises has been completed three or four exercises in laying out work by means of a square, such as finding the length and the cuts of the different kinds of rafters, the length and the cuts of a trestle leg, hopper joints, truss joints and members, and stair horses may be given. These will serve to fix his knowledge of practical geometry and assist him in mill work.

The course of wood work at the bench should be followed by a course in wood turning. The different methods and devices employed on the lathe should be taught preparatory to pattern making. The instructor should take apart the different woodworking machines and have the students re-assemble them and adjust the machines under his strict supervision, being careful to explain wherein lies the danger in the use of the several machines.

During the second year blacksmithing should be taught. The drawings of the exercises wherever required should be made on a piece of sheet iron or on the forge-hood free-hand, with the necessary dimensions, as a foreman gives instructions to a blacksmith. Working from blue prints and sketches should also

14

be taught. Where curved pieces are required they should be tested to a templet or full-sized drawings made with chalk on a piece of sheet iron. In the blacksmith shop, the first exercises should first be wrought out of lead and then out of iron. My faith in this method is founded upon experience and strengthened by observation. Wood or cold iron will wait any desired length of time, while the student stops to think what he is to do and how he is to proceed, but hot iron will not wait and he must think and strike simultaneously while the iron is hot.

This shows the advantage of lead over iron; with lead the student can consider how to proceed while all the particulars of holding the metal, striking, and the different operations on the anvil are first learned. Lead acts under the hammer very much like hot iron, and all the operations except welding can be performed with it on the anvil. Furthermore lead is more economical than iron, it saves time in not having to wait for heating, it saves material and coal, (the lead can be melted into new bars,) and it allows of better workmanship—a property of great satisfaction and inspiration to the student. Among the most essential things to be impressed upon the student's mind is the management of the fire and the working of the iron at the proper heat required for the operation in hand. The series of carefully graded exercises should include drawing, bending, upsetting, punching, welding, tempering, and brazing; then some exercises involving as many of the foregoing in different combinations giving more difficult forms. The final exercises should be a complete set of machine and hand tools for use in machine shop, several blacksmith's tools and some forging for a machine to be built. The instructor should explain the effect of the different operations on the metal, showing the injury done by a different method. The properties of the different current commercial brands of iron and steel and their special uses and how to identify them should be clearly taught. The amount of material necessary to make a given piece of work without entailing excessive waste should ever be explained, thereby encouraging the economy of material in manufacturing. Steam hammer work should follow anvil work with reference to the designing of tools, dies, and formers to facilitate uniformity and economy of production. The work on a horizontal press or "bull dozer" would be a profitable study in relation to the formers required for bending iron into the various shapes which can be done more uniformly and cheaper on the "bull dozer" than on the anvil. One object sought should be to acquaint the student with the various labor and time saving appliances used in a large shop. The coal and iron used in the blacksmith shop should be of good quality. In some cases Norway iron should be used, for poor iron and poor coal will invariably lead to poor results in the hands of beginners and this in time leads to discouragement, a condition which it should be the unceasing aim of an institution to avoid.

In the third year sheet iron work should be taught. This further develops the student's ability for laying

out work and assists in fixing his knowledge of inter-
sections and development of surfaces. This should be
followed by pattern-making and molding. The student
has acquired the use of the tools from previous training.

The fourth year should be devoted to the machine
shop and machinery. Each student should be carefully and thoroughly drilled in the use, care, and management of the boiler and engine, together with all
their accessories. In the machine shop the students
should work entirely from blue prints or from drawings
carefully prepared in the drafting room. The series of
exercises should involve at the bench the use of the vise,
hammer, chisel, file, scraper, hand dies, taps, and
reamers; the use of lathe, drill-press, planer, shaper, and
milling machine; and also the use of the various tools
and devices to perform at least the ordinary operations
on these separate machines. The care of the tool room,
and the construction and use of templets, gigs, and
measuring apparatus should be a leading feature. The
final exercise should be a machine, device or engine,
often the product of the combined labor of several
students, all to be made in accordance with blue
prints carefully prepared by the students and corrected
by the instructor. One of the chief aims of this shop is
to enable a student to design a machine or part of a
machine, or to design work to be done on such a
machine according to scientific principles, expeditiously, and at a minimum cost.

The instructors in the various shops should be
careful to explain the successive steps, they should

state the principles involved, the nature of the properties of the material, used and the object aimed at.
It should be his constant aim to show the best methods
and he should explain their advantages over others. All
the tools or machines involved should be carefully explained, the manner of their working, their particular
construction, and above all their care and abuse. The
instructor should explain the object of each exercise,
whether merely of educational value or of educational and practical value, and when used to cultivate
the student's habit and power of observation. All
explanations should be made by the instructor while
performing the work on the exercise. The students
should have all the practice possible in handling
engines of different types, in taking them and other
machines apart, examining the details of internal
construction, and reassembling the parts and adjusting
the machine with exactness.

In conjunction with the shop work, visits should
be made to reputable shops using the latest and most
improved machinery in order to further familiarize the
students with current commercial work on a large
scale. Particular attention should be directed to the
form and size of such machine parts as can readily be
constructed in the various shops; to the time required
to perform each operation, in order to give the students some idea of the cost; to the dimensions required
by the mechanic; and to the various tools and devices for
expediting work and insuring uniformity and accuracy.
This will acquaint the student with the knowledge

required by the foreman and superintendent. Each
student should make a sketch and a study of some
special device, or machine, or operation, and upon his
return should write a full description including the
necessary sketches. These reports, having been corrected by the instructor, should be read and explained
in the presence of the whole class so that each student
gets the advantage, as far as possible, of another student's study.

The instruction given in the various shops should
be given by trained instructors who have had the
advantages of a good liberal education. A mere
mechanic will not and can not fill the bill; he is too
apt to teach the students in the way in which he himself
learned his trade. Without a knowledge of underlying principles the instructor can not give full explanations, and he will expect the students to learn by
watching him perform the work. The leading idea in
all the instruction should be to teach principles rather
than produce objects of commercial value; for the
greatest progress can be made in a given time by
adhering to this method. The students should perform their work in the shops at regular hours, as a
class, the same as any class-room work, and not be
allowed to put in only the required number of hours
at the side of a so-called foreman or other mechanic in
a shop where commercial articles are made. Neither
should they be allowed or required to learn what they
can at the side of a mechanic and perform all the
chores that are commonly done by an ordinary appren-

tice, for a student's time is too valuable. The student should be ever learning new principles and methods from the time he enters the workshop until he leaves it. This naturally raises the question of the time which a student can profitably spend in a shop at each period. In some institutions the students are required or allowed to spend one or more entire days in the workshop beside a mechanic who is not an instructor. This is a bad and profitless practice and should not be tolerated.

At other places students are allowed to work from four to five hours at a time, perhaps every afternoon, either to remove a condition or to get some advanced shop work out of the way in order to take up something else. This overtaxing of the time and endurance of a student should not be allowed. I do not mean to discountenance or discourage extending the shop privileges to students out of hours, to do some work for themselves, for they are then only applying what they haved learned and are not under instruction to learn new principles or methods.

The writer seriously doubts the advisability of extending the period of time for shop work beyond two hours. He has tried four hours and three hours and found that the most and best work was accomplished in periods of two hours, covering the same total number of hours per term. The number of periods per week should be at least three. When a student's mind becomes weary and his attention flags, he is no longer alert in learning a new principle, and it is not profitable for him to spend any more time in that branch.

The writer believes that a thorough course in shop work in our engineering schools as here described will facilitate that adaptation of a technical education to the industrial arts which is so much needed, and which will widen the field of usefulness for the graduates of our schools of engineering.

MECHANICAL LABORATORY WORK AT AMES, IOWA.

By G. W. BISSELL,

Professor of Mechanical Engineering, Iowa Agricultural College, Ames, Iowa.

The mechanical laboratory of the Department of Mechanical Engineering of the Iowa State College of Agriculture and Mechanic Arts is located in a one story brick building, 36x78, substantially built and designed primarily as a power house which purpose it now serves together with its present use as a laboratory. It contains two rooms known as the engine and boiler rooms respectively.

The power equipment consists of a 50 H. P. Babcock & Wilcox boiler with feed-pump, inspirator, exhaust steam injector and the usual appliances for testing purposes, including pyrometer, flue-gas collector, draught gauge, calorimeter, etc.; an 8x24 Harris-Corliss engine, running at 100 revolutions per minute, used chiefly to furnish power for the shops, but equipped with calorimeter, indicator connections, tachometer, stroke counter, etc., for experimental use; and an "Otto" gasoline-gas gas-engine of 12 I. H. P., lately installed and intended, on account of economy of fuel and attendance to replace the Corliss engine for power; it will be fitted up for testing. The additional equipment of the building is used for educational purposes

217

and consists, in addition to four dynamos and two motors belonging to the Department of Electrical Engineering, and cement testing apparatus belonging to the Department of Civil Engineering, of the following: one 5 H. P. throttling steam engine with Prony brake, used for valve setting problems—sometimes loaded with a number 7 Sturtevant fan fitted with blast gates for varying the power required to drive it; one Westinghouse air-brake pump; one Blake air pump, 6 & 8x6, fitted for pump duty tests; one 50,000 pounds Olsen testing machine, one 2,000 pounds transverse testing machine; one cradle dynamometer; one Morin dynamometer especially adapted for measuring the force exerted on lathe tools; one Pelton water motor; one weir tank; one 50 H. P. Wheeler surface condenser into which the steam from any engine or pump can be discharged at will; five steam engine indicators; flue-gas analysis apparatus; viscosity apparatus; and an assortment of gauges, thermometers, speed indicators, micrometers, platform scales, etc., together with the necessary apparatus for the calibration of instruments.

The course of instruction in laboratory work begins with the second half of the Junior year. The time required is one afternoon per week for sixteen weeks. In the first term of the Senior year two afternoons, and in the second term one afternoon per week for sixteen weeks complete the course. Thesis work gives additional opportunity.

The following experiments are performed by the members of the Junior class: Tension, transverse and compression tests of materials, calibration of steam gauges, indicator springs and thermometers, calibration of weir and water meter, comparison of barrel, separating and throttling calorimeters, viscosity, specific gravity and flashing and burning points of lubricants, steam engine valve setting with indicator practice and use of Prony brake, and flue-gas analysis.

The Seniors in the first term are assigned the following: Efficiency tests of steam engine, steam boiler, injector, gas engine, hydraulic ram and water motor, duty test of steam pump, cradle dynamometer test of small fan blower, tests of large fan blower, test of college electric lighting plant and tests with Morin lathe dynamometer.

In the second term the Seniors conduct special experiments which include tests of steam heating plants, the college pumping station, and such commercial work as can be used to educational advantage.

The students work in groups of two and three in taking data and working up numerical results, but prepare separate reports. Carpenter's Text-Book of Experimental Engineering is used as a basis of the conduct of all the experiments in the curriculum but is supplemented with notes as may be necessary. For guidance in writing reports students are advised to group the subject-matter under heads as follows:

1. Object of the experiment.

2. Method to be employed in attaining it. Under this head is placed the derivation of the fundamental formulæ for the experiment.

3. Description of the apparatus, principal and accessory with the aid of sketches where desirable.

4. Describe the experiment—every operation having direct or indirect bearing on the results.

5. Give numerical data.

6. Derive results.

7. Draw conclusions.

Printed blanks are furnished for most of the experiments. All work is done during hours of instruction and under supervision. For the current year the Seniors, nineteen in number, are in charge of the writer and the Juniors, thirty-five in number, are handled in two sections by Assistant Professor W. H. Meeker.

For obvious reasons thesis work is largely experimental and calls into play the resources of the laboratory. Some of the work done this year is on original lines, a few of the subjects being: "Effect of Variation of Clearance Volume upon the Economy of a Small Throttling Steam Engine." "Autographic Record of the Stresses in the Punching of Materials." "Loss of Head by the Flow of Water under Small Head through Horizontal Swing Check Valves." "The Minor Losses in an Electric Lighting Plant."

THE EQUIPMENT FOR ELECTRICAL ENGINEERING LABORATORIES.

By DUGALD. C. JACKSON,

Professor of Electrical Engineering, University of Wisconsin, Madison, Wisconsin.

When this subject was assigned to me by your secretary, it was suggested that a discussion of the *ideal* equipment for the electrical engineering laboratories of engineering schools was not wanted. The discussion should be limited to the special electrical equipment that is *necessary* for use in engineering schools having a course in electrical engineering, in order that proper instruction in their specialty may be given to the undergraduate electrical engineering students. This seems to be a reasonable limitation, and, in fact, an essential one, if the discussion is to be of value. I doubt if anyone is able, with our present knowledge of the requirements, to outline the true ideal equipment for electrical engineering laboratories. We can all tell what we now consider an ideal equipment, but in a few years it may be far from ideal. We are, however, faced by a positive condition which demands a first-class working equipment in the electrical engineering laboratories. *How can the electrical laboratory equipment be made most effective*, is a question for solution; *how much and what equipment is*

221

required to justify the demand in the instruction, is another question which confronts us.

As we are dealing with *electrical engineering* laboratories, we will assume that they are attached to an engineering department, over which a professor of electrical engineering presides. This department is therefore separated entirely from that of physics, and is coordinate with the other departments of engineering. The practice, at present common in many colleges, of placing the electrical engineering course with its class-room and laboratory instruction under the direction of the professor of physics is just as absurd as it would be to devolve the instruction in advanced thermodynamics and the design and construction of steam engines upon that professor. It is equally absurd for the professor of electrical engineering to attempt to teach physics. Proper instruction in physics is essential to the student who is pursuing a course in electrical engineering, and the physics department must be well controlled and supplied with an excellent independent laboratory equipment. In addition to the ordinary general course in physics, the electrical engineering student should gain, during his course in physics, a common sense grasp of the elementary notions of electricity and magnetism, and of the "all pervading law of Ohm." The latter can be properly enforced in the laboratory by placing in the student's hands ordinary electrical instruments, such as galvanometers, bridges, voltameters, amperemeters, voltmeters, etc. Before beginning his work in electrical engineering, the student's knowledge of Ohm's law and its common results should have become almost instinctive. The student should also become acquainted with the theory and practice of photometry. This should all be included in the physics course, and the equipment required for the instruction should be controlled by that department. Much of this equipment is quite different from that required by the electrical engineering department, but some of it is useful to both departments, and where students are not too numerous it may be used to some degree in common.

With due regard for his preparation, it seems best to arrange the professional electrical studies, for the average electrical engineering student, in four divisions, in three of which laboratory work must be made an integral part of the instruction in a satisfactory course. The divisions are:

First. Electromagnetism and its applications to special uses, with special reference to dynamos and motors.

Second. Electrochemistry (including primary and secondary batteries) and electrometallurgy.

Third. Alternating currents and alternating current machinery, including alternators, transformers, motors, condensers, etc.

Fourth. The special applications of the preceding divisions in electric light, power, railway, mining, telegraph, telephone and other types of plants. These divisions are of about equal real moment, but the last must be allotted more time in the lecture room than either of the others, while little or no laboratory time or equipment need be directly allotted to it. The laboratory equipment for the first three divisions only, therefore, need be discussed. The equipment that can be made useful in the fourth division is usually of an illustrative nature, and must be a matter of slow growth and collection as the electrical industries progress, rather than of immediate selection.

While higher mathematics is a useful aid in each of the divisions, its limitations as an agent must be carefully shown in the class room and laboratory. For the purpose of educating the judgment of a student, and fully defining the limitations of theories and mathematical deductions, the laboratory is indispensable, and the equipment should be selected with this object continually in view. The equipment should be so selected, and be of such amount, that as much as one half of the total time spent by the student under the direct instruction of the professors of electrical engineering may be profitably devoted to the laboratory. The equipment, moreover, should be such that the work may, as far as possible, deal with fixed commercial instruments and machinery, and actually follow the methods of testing and research used in practice. The first three divisions should each be properly represented by a commercial laboratory equipment, in which every principle and operation is illustrated and made useful in every day instruction, under the direction of a man who has had experience in similar commercial work, after having had a thorough theoretical training in preparation for his work.   The laboratory work must always go hand in hand with that of the class-room illustrating and strengthening the latter.   *In this way the equipment may be made most effective in the work for which it is designed.*

We will call the laboratory apparatus for the three divisions, the dynamo equipment, the electrolysis equipment, and the alternating current equipment, and consider the essential features of each separately in the consecutive order given, since that is probably the best order for the student to take them up.   It is well to remember at this point that engineering measurements are almost invariably quantitative, and that the results of an engineering investigation which bear fruit are also nearly always quantitative. It is seldom that purely qualitative results may be accepted as satisfactory in engineering work. Therefore, the laboratory equipment must be capable of doing quantitative work which is, in general, of satisfactory commercial accuracy, and which may be made rigorously exact, for purposes of exact investigation.

The first work that a student should take up upon entering the dynamo laboratory is a study of the methods of testing the magnetic qualities of iron.   In this work, testing by magnetometer, ballistic galvanometer, and traction should be used.   The apparatus for the magnetometric method is very simple, and is generally at hand, as almost any reflecting galvanometer may be used for the magnetometer.   For the

15

ballistic method, a magnetizing yoke and proper test pieces should be supplied, as well as a satisfactory galvanometer. Arrangements for determining the constant of the galvanometer and proper current measuring instruments must be supplied. For the latter, Weston amperemeters with proper scales and in good condition are entirely satisfactory. To determine the constant of the galvanometer, several methods requiring their own special but simple apparatus should be used. The special apparatus for the traction method of testing iron, consists simply of a magnetizing yoke with its test pieces. Here again an amperemeter is required. Indeed, a good supply of first-class amperemeters and voltmeters is of the utmost importance in the electrical laboratory. These should be selected so that various types are represented, but for general use with continuous currents, the Weston instruments are most satisfactory, and should, therefore, be in a majority. Amperemeters with scales from 0 to 1, 0 to 5, 0 to 15 and 0 to 50; voltmeters with scales from 0 to 5, 0 to 60, 0 to 150, 0 to 600, and milli-amperemeters and milli-voltmeters, are indispensable. The actual number of instruments required for use must depend, of course, upon the number of students who work in the laboratory. In addition to the standard methods of testing iron which have been mentioned, there are several special devices, constructed by different instrument makers, which are of some advantage in instruction, but which should not be purchased until all the essential equipment is supplied. Such are the iron tester, using a bismuth spiral, made by Hartmann & Braun; the Siemens & Halske device, and magnetic bridge arrangements.   It is not a bad plan to have various test pieces, for use with the ballistic method, made from cast and wrought iron and cast steel of known chemical composition and degrees of annealing. Those are most convenient when made in the form of bars to be used with the magnetizing yoke, or may be rings. By using various specimens for the class testing a very good idea of the differences to be found in different bars is given to the students.   In this work the student should always be required to plot both the curve of magnetization and the permeability curve of his specimen, as well as to fully report his process.

After the testing for simple magnetic quality comes the measurement of the hysteresis loss and its companion, coercive force.   For this purpose apparatus for the magnetometric method with slowly varied continuous currents should be provided.   This may be similar to the apparatus for the magnetometric method of testing the magnetic quality of iron. If a sufficiently small number of students work in the laboratory the same apparatus may be used.   In addition to the magnetometric apparatus a couple of devices for testing hysteresis loss by alternating currents should be supplied.   These methods are being improved so rapidly that it is impossible to say which are the best for fixed use in laboratory instruction, and much must be left to individual choice.   Here, again, there are various excellent illustrative but unessential pieces of apparatus, such as the Ewing curve tracer, which should only be purchased after the essential apparatus is fully supplied, or which really belong in the field of the department of physics.

The writer makes all his laboratory work follow quite closely to his class-room instruction, and, therefore, makes the work in the dynamo laboratory follow the plan of his text-book on Electromagnetism and Continuous Current Dynamos. Consequently, after the students have completed their work in magnetic testing, they begin a study of the dynamo. The dynamo room must fulfill two objects: *First.* To furnish current for experimental work which can not be conveniently furnished from the laboratory lighting circuits. *Second.* To supply machines for laboratory testing. For the first purpose, dynamos of various pressures, driven by a steady power, must be arranged so that they may be operated upon laboratory circuits as desired. This requires that a good automatic engine which is used for no other purpose be used to drive the dynamos, and that the machines be systematically wired to suitable distributing switchboards. A good storage battery of sixty or seventy cells is a most useful adjunct in this connection. For the second purpose it is desirable to have continuous current machines of comparatively small size—from two to ten kilowatts capacity. Such machines are readily handled, and are not readily injured, when injured are inexpensive to repair, and laboratory testing with the machine under load is always possible, while it would frequently not be so with larger machines. When machines of smaller capacity than two kilowatts are habitually used for laboratory testing, the students do not gain a proper knowledge of the right methods of handling dynamos. A dynamo laboratory in which a fair number of students work needs a number of such machines, which should be carefully chosen to represent the widest range of types and styles of construction. All these machines should be so arranged that they may be operated either as dynamos driven from a constant speed prime mover, or as electric motors.

Upon entering the dynamo room for work, the student receives systematic instruction in handling the dynamos so that his work may proceed without danger to himself or to the machines. Determinations of the strength of the fields of dynamos when the armature windings are known, are then made from measured speeds and pressures. This requires no instruments, except voltmeters and ordinary speed counters. Following this, comes the measurement of leakage by ballistic methods, and a comparison of the measured results with results calculated from the dimensions of the machines. This gives the student an idea of the large percentage of guess which must enter into such calculations. Tests of the magnetic circuits of dynamos, the regulation of dynamos and motors with different forms of field windings, effect of load on lead, characteristic curves, etc., all require the all-useful amperemeters, voltmeters and speed counters, which, as already said, must be on hand in sufficient number and with a sufficient variety of range.

Efficiency testing requires a more varied equipment, yet, in this, the amperemeter and voltmeter hold prominent places. In stray power tests and determinations of the losses by hysteresis and foucault currents in armature cores, amperemeters, voltmeters and speed counters supply all the information that is needed. Efficiency tests should always be made by stray power, live load, and dead load methods. For the latter, convenient variable liquid, wire, and lamp loads should be liberally supplied. One or more first class bridges must be at hand for measuring resistances, and standard instruments, such as Kelvin balances or Weston laboratory instruments, should be conveniently arranged for the calibration of the portable instruments. Stationary devices for the measurements of currents, such as a good calorimeter and a galvanometer arranged for use in the so called Vienna method, should be properly mounted.

In brief, the essential apparatus for the dynamo laboratory are: Appliances for testing the magnetic and hysteretic qualities of irons, together with the necessary test pieces, galvanometers (plain and ballistic), standard solenoids, earth coil, condenser, etc.; set of small continuous current dynamos and motors, representing various types for laboratory testing; goodly supply of first-class amperemeters and voltmeters, of various types, with standards for use in calibration; instruments for measuring currents by different methods, bridges and special instruments for measuring resistances; various arrangements for loading dynamos and motors, and testing their speed; a number of portable storage cells to furnish steady currents, which can not be conveniently supplied from the lighting circuits and the dynamo room.

The electrolysis and alternating current equipments will not be taken up in so great detail. The former requires for the fundamental work a number of standard cells, silver and copper voltameters, fine galvanometers, balances for weighing, accurate resistance boxes, devices for measuring the resistance and specific inductive capacity of electrolytes, etc. For the advanced work is required a goodly supply of primary and secondary cells of various types which may be used in examination and testing, a well equipped set of two or three good sized electrodeposition vats, and arrangements for heating by the arc, with low pressure dynamos, and the amperemeters, voltmeters and hydrometers which are necessary to carry out work on a considerable scale in electrodeposition and electrometallurgy. The electrolysis laboratory has received comparatively little attention in American engineering schools, but its importance makes it deserve an allotment of its full share of equipment and instructional force. There seems to be a time coming when the electrolysis equipment will be made equal in importance to the dynamo equipment in all first-class engineering schools.

The alternating current is receiving a preponderating amount of attention in many colleges, sometimes on account of the numerous striking and beautiful experiments which may be performed by the alternating current, and sometimes because of the attractive field of speculation which is opened by the use of polyphase currents in power transmission. It is not belittling the enormous value of the alternating current to say that the equipment of each of the divisions should be given equal consideration, and no one should be given undue prominence to the injury of the others.

The effects of self-inductance, mutual inductance, and capacity are at the root of a large proportion of alternating current phenomena, and the fundamental alternating current equipment should be arranged for the study and measurement of inductances. It would be a long step in advance if electrical engineering graduates were given as lively and practical a working conception of inductance and reactance as that usually given them of Ohm's law and its results. For studying inductances in the laboratory, various coreless solenoids are needed which may be used for experimental and mathematical determinations of the coefficients. The methods of measuring the inductances of solenoids should include the direct measurement of self-inductances by amperemeter and voltmeter, using an alternating current of known frequency, the comparison of self-inductances with each other and with mutual inductances, the direct measurement of mutual inductances by amperemeter and ballistic galvanometer, the comparison of self and mutual inductances with capacities, etc. After these measurements have been carried out by the student, using the simplest known methods, the measurements should be extended to include circuits with iron cores. These measurements require comparatively simple apparatus, but it should be on hand in sufficient amount. Carefully wound long and short solenoids of known constants, condensers of known capacity, bridges, alternating current amperemeters and voltmeters, a sensitive ballistic galvanometer and a secohmmeter are essential for the work. A very convenient instrument for use in this branch of the work is a variable standard inductance, but its place may be filled by solenoids of known constants.

After inductances, the alternator and single phase currents are in logical line. One or more small alternators are needed for this work, which may be carried out in much the same way as that relating to continuous current dynamos. The work, however, requires the introduction of two new instruments, the wattmeter and a device for taking alternating current curves. Wattmeters are difficult instruments to get in satisfactory form, but probably the Weston form is the best for general laboratory use, though a split dynamometer is a good instrument to have at hand. There is a considerable number of schemes for the tracing of alternating current and pressure curves, and more than one should be put into use. The relations of the current and pressure waves in circuits should be studied from the curves, as well as from measurements of the true and apparent energy in the circuits. This study is finally applied to transformers, a number of which, representing various types of construction, should be in the equipment.

The laboratory equipment to be supplied for the study of polyphase current effects is, as yet, an open question. It seems desirable to have at hand polyphase currents of variable frequency, and induction motors representing the different types of construction. The polyphase currents of variable frequency may be most readily gained by means of a rotary converter.

The essential requirements of any first-class, undergraduate course in electrical engineering are covered by the equipment which has been enumerated, or which is necessary to carry out the experiments that have been mentioned. There is much more work which must be carried out in the laboratory, such as testing the insulation resistance of wires, measuring the earth's magnetic force, accurately comparing standard resistances by divided wire bridge, etc., etc., but the equipment for this is found in that already enumerated, or has its place amongst the apparatus of a well equipped physics department. There is much apparatus which is entirely unnecessary for both the physics and electrical engineering departments to possess. In fact, the work just mentioned, with the exception of insulation testing, really belongs in the instruction given by the professor of physics, and, as intimated earlier, it is well for the laboratories to be quite independent of each other. There is a vast amount of work carried on by undergraduates in the preparation of theses, which is often in the nature of excellent investigation. Some of this requires special apparatus, and the extent to which it should be supplied from laboratory funds is an open question. Where the investigation does not require special apparatus, or where the work is in the nature of a test of commercial apparatus or an operating plant, the equipment that is required will nearly always be found amongst that supplied for the regular course, if the latter has been properly designed.

In determining the course of study to be pursued by the electrical engineering undergraduate, upon which the course of laboratory work must depend, it must unfortunately be remembered that he can not confine his attention to electrical engineering during the whole time given to professional work in his course. He must gain an elementary but practical knowledge of thermodynamics and hydraulics, with an efficient knowledge of their application in steam and water-power plants. He must also get a common sense knowledge of the principles underlying the design, manufacture and selection of machinery. In fact, he must receive a good working knowledge of the problems of the mechanical engineer. Laboratory courses are likely to fail when not properly balanced, on account of a failure to educate the *common sense or judgment* of students, and the students leave the college without having gained an all-round capacity for practical work and research, which is necessary to put them in a fair way to become useful engineers. The knowledge of the electrical engineer must be based on the honest, well tried mechanical laws, and he must go into a study of all that will aid most in putting him in the way to make a thorough electrical *engineer*.

TEACHING MACHINE DESIGN.

By JOHN H. BARR,

Assistant Professor of Mechanical Engineering, Sibley College, Cornell
University, Ithaca, N. Y.

Every specialist is prone to believe that his own
particular specialty is surrounded by greater difficulties
than is any other line of work, and that it requires more
persistence and genius in the attainment of success.
Probably every teacher of machine design (unless he
happens to teach other subjects as well), will admit
that no other branch in the mechanical engineering
curriculum is so hard to handle satisfactorily.

What is the legitimate aim of such a course?   It is
not to turn out full fledged Bements, Reynolds, Sweets,
and Leavitts; but only to develop the talent, and stimulate the growth of the young men who fall into our
hands; to turn them out with some appreciation of the
conditions they are to meet; to familiarize them—as
far as possible—with the usual methods of design and
construction; and above all to impress upon them the
practical limitations of these methods. We can do
these things but incompletely, but we should be able to
do enough at least in this way to accustom the student
to study such matters for himself. We may be
reminded that only a few of the students with whom we
come in contact have real talent for design, or will ever

236

be thrown into this branch of engineering as a specialty. This is true, and to an increasing degree as the industries of the land become more highly specialized. The number who operate machinery and direct construction is increasing, relatively to the number engaged in designing; and right here with this larger class we have a duty quite as important, and certainly no less arduous, than that imposed in the development of the talent for design in a more gifted man. It is almost as necessary for the man who selects, inspects, erects and operates machinery to appreciate the good and bad points of design as it is for the designer himself to do so. It is these embryo operating engineers to whom we must devote ourselves, largely, both because of their greater numbers, and of their smaller capacity, for the particular kind of work we have under discussion. The genius will discover the errors of our precepts before he has traveled far in the regions of design. The student who drops conscious attention to such matters when he receives a passing mark, is less apt to detect the flaws in our dogma; and the impressions we leave on him, (slight though they may be), are more liable to be permanent, and should be corrected.

It may be urged that the genius needs no coaching to bring out his talent; as to this I am sure I can do no better than to quote from a most admirable paper by Mr. John T. Hawkins, printed in the Transactions of the American Society of Mechanical Engineers (Vol. VIII, page 458). This paper entitled: "The Education of Intuition in Machine Designing," is one that every teacher should read, every teacher of engineering should read it twice, and every teacher of machine design would do well to read it about once a year. Mr. Hawkins says: "I take it that the faculty in man which may perhaps be properly defined as mechanical intuition is as much dependent upon education for its full development as a natural tendency to be musical, or an innate talent for painting, or any of the fine arts. It is well understood that, no matter how great a prodigy a boy may show himself to be in either of the directions last mentioned, the cultivation of his particular tendency is indispensable to any marked success in the application of it; and it is doubtless true that a natural mechanic may be developed by cultivation to as great an extent as a natural musician, or artist, or orator; and that, without development, either of them must necessarily be deficient."

On the other hand, it is generally conceded that a cultivation of one's taste for literature, fine art, or music, is not wasted, even if the subject lacks the genius for creation in these liberal branches. Likewise, the appreciation of good design is not a useless acquirement with the engineer in whatever line of work he may engage. We who are especially concerned with this part of the training of engineers must then keep in mind the requirements of two more or less distinct classes of students; and when we reflect on the multitudinous branches into which these young men will soon find their way, it may seem difficult to plan a course which will be suited to the needs of any two individuals. Of course we cannot know the future of any student, and it is well that we cannot. An attempt to shape our course to the detailed requirements of his professional career would doubtless be unsuccessful, and if successful (in the sense of meeting his problems for him in advance, and sending him out charged for a specific duty), our effort would result in that dismal failure, a man narrow and without self-reliance. On the other hand, we do not feel that the sole object of our instruction is "mental discipline," else why have we broken away from Greek, Latin, and mathematics? That diet has been the nourishment in youth of some of our most eminent engineers; a much less nutritive mental food has been the lot of many others; yet, among both of these classes are to be found the staunchest friends of our present more specialized training. The magnificent success of the mechanical laboratory—the quintessence of special training—in our technical courses is the best of evidence that our newer education is in the right direction. The laboratory is one of the foremost (perhaps the first) of the essential elements of an ideal training in design; for nothing so impresses a man with the good and bad features of a machine (especially the latter) as the actual manipulation of that machine.

In the first place, a word as to the shaping of our course to the needs of the two classes of students whom we have to teach. The difficulties of giving each his proper attention are not so great as might seem at first glance. *First.* Because the man who is to become the operator, superintendent, or inspector of machinery needs quite as much drill during his undergraduate course in the details of design as does the future designer, for the latter is to continue through life the study of design as a specialty, while the former gets a larger proportion of his training in this branch in the college course. *Second.* The instruction in the drawing rooms, to which the lectures should be merely auxiliary, is largely individual, and it can be adapted to the capacity and needs of each student. The man with talent for design will himself, under the influence of proper suggestion from the instructor, do collateral reading and study of perhaps even more value to him than the formal instruction.

In the establishment of a plant for the utilization of a material it is in order to first become familiar with this substance, its form, properties, and capabilities; then to determine what may be made of it; and finally to consider ways and means for conversion of the raw material into a marketable product. The finished product of the rolling mill is the raw material of the bridge and the steam boiler. We are operating a mill the finished product of which is the raw material for making engineers. How can we put the most valuable product on the market, final efficiency being the true measure of value? What shall we teach and how shall we teach it?

Let us begin by noting a few of the requirements which a machine designer must have constantly before him in the creation of a machine. The first requisite of a machine is utility. The machine must be adapted to perform a specific operation, or series of operations, in an accurate, economical manner; and the general primary requirements of a machine may be enumerated as: adaptability; strength and stiffness; economy in construction and operation. A secondary consideration, but one that may largely affect the commercial success of a permanent construction of the better class,—is appearance.

The adjoining outline indicates something of the scope of these general considerations, though it could no doubt be amplified.

GENERAL CONSIDERATIONS IN MACHINE DESIGN.

ADAPTABILITY.
- (a) Kinematics.
- (b) Character, quality, and quantity of product.
- (c) Simplicity.
- (d) Safety.
- (e) Convenience.
  - Operating.
  - Adjusting.
  - Repairs.
- (f) Facility of transportation.

STRENGTH, RIGIDITY, AND STABILITY.
- (a) Resistance to breaking (strength).
- (b) Resistance to undue distortion (stiffness).
- (c) Mass of stationary and moving parts (steadiness of running).

ECONOMY.
- (a) Cost of construction.
- (b) Cost of output.
- (c) Durability.
- (d) Cost of maintenance (oper'n, rep'rs).
- (e) Convenience of adjustment and repairs.
  - } Friction, etc.

APPEARANCE.
- (a) Proportion (form and outline).
- (b) Harmony of parts.
- (c) Finish.

The relative importance of the above named primary requirements varies in different cases, and they are usually mutually interdependent. In nearly every instance, the first consideration is adaptation of the machine to some more or less specific duty. Thus a planer is designed to receive energy from a rotating

16

shaft, and transfer it in such manner that a proper reciprocating motion is given to the platen under the tool. This is the first element, furthermore, the required feeds must be provided. Next in importance to adaptability, in this example, and really a function of it, is strength, or rather stiffness, for the question of rigidity commonly overshadows that of mere strength, especially in machine tools. Of course every part must be strong enough to prevent rupture in any ordinary contingency, but in general, a machine member is ruined if it takes any appreciable permanent distortion or set. In machine tools the maximum permissible distortion in use is ordinarily much less than that corresponding to the elastic limit of the material, otherwise the accuracy of the product is seriously impaired. In such machines as presses and punches, strength is a more prominent element; in a fine lathe it need hardly receive a thought, so essential is rigidity and limitation of wear of the bearing surfaces. Economy may almost be said to be the one essential of a machine, for adaptability and strength and stiffness are functions of economy. A machine to be economical must be adapted to its service, and it must be stiff and strong enough to perform its duties well or economy is not attained in the highest degree. If it is so nicely adapted to its work that the required service is performed in a direct and "mechanical" manner, one element of economy is secured. Stiffness not only affects the accuracy of the product, but springing of the running parts of their guides frequently results in loss through friction and in serious wear. Other elements of economy are cost of construction, of adjustments, and of repairs. Without attempting the impossible task of treating this subject at all completely in a single paper, it may be worth while to touch upon a few of the many considerations involved in the design of any important machine.

Two things to be kept in mind by the designer may be emphasized—the number of machines, or pieces of a similar form, to be made; and the number of operations to be performed by the machine. If, for instance, the machine is a temporary thing to be used but once, or at most but a few times, and if but a single machine is to be built, castings may be replaced (where they would normally be used) by forgings or even by timber; or the ideal form of casting may be sacrificed economically, to reduce the cost of pattern making. An expert moulder can turn out remarkable castings from a few strips, blocks, and sweeps; and core boxes and split patterns are expensive. If, on the other hand, a very large number of similar pieces are to be made, take care of the pennies in the foundry, and the dollars will take care of themselves in the pattern shop. A properly arranged course in the school shops in moulding and pattern making (and I would prefer taking them up in this order) is a most valuable adjunct to a course in design. I believe that these courses should be arranged with this as the predominating object in a school of mechanical engineering, and that the departments of design and shop work should be in close touch. I am heartily in sympathy with the work of the manual training schools, and hope that the time may come when all students entering our engineering institutions shall come with such a training as these schools give; but I look upon the shops of the professional school as a part of the professional training, and as such, their methods should be radically different from those of the manual training school. It is not the intention of this remark to reflect upon the present shop courses of the engineering institutions of to-day, for until the elements of manual training can be required for entrance to them, this instruction must be given during the valuable time of the college course. When this condition has passed away the professional school will take a great stride toward a higher plane.

The designer must study processes of construction; he must be familiar with the methods and limitations of the shops, especially with the relative efficiencies of the equipments and workmen in the various departments of his own works. These so-called "practical" matters (and they certainly are practical) play the star parts in actual designing of the most successful works. Not only is it important that the capabilities of the shop equipment be kept before the designer but he must as constantly have in mind the duties of the machine which he is designing, and especially the range of its operations. In the manufacture of guns, sewing machines, and typewriters, we see many tools employed, year in and year out, upon a single piece, perhaps performing but one operation in continual succession. These are special tools in the strict sense. In a jobbing shop we find the universal milling machine used on the greatest variety of work; this is a general machine in the fullest degree. Other machines occupy the intermediate territory. In the special machine it is essential that the time and care required in maintaining the standard product be reduced to the minimum; and permanence of adjustment is the desideratum. In the more general machines facility of change of certain parts (with reasonable security) is of more importance, relatively. In the proper compromise of these two requirements, the designer often has scope for the exercise of his best judgment. Changes that are to be frequently made must be made conveniently, without the sacrifice of good mechanical construction; while the maintenance of alignment, position, and form should be insured for those parts which are normally changed only as repairs become necessary. Other practical matters to be looked after are the use of standard forms of materials wherever this is possible, clearance for moving parts; and provisions for taking up wear. Many other items might be enumerated which demand the closest study on the part of the designer, but the few suggestions, taken somewhat at random, must suffice, and the teacher who has the qualifications for giving instruction in machine design will readily supply the deficiencies.

The existing text-books on machine design, with but few exceptions, may be classed as treatises on theoretical mechanics. Some of them are excellent in the field which they cover, but are far from comprehensive as works on design. In the treatises now available, little attention is given to the real essentials of machine construction, save the one secondary element of strength of materials. In the great majority of machines the proper office of this element is that of a check, simply. It bears somewhat the same relation to the complete machine that the inspection and insurance of the boiler bear to the steam plant. In certain instances this insurance is more important than in others. I believe, emphatically, in teaching this subject of strength as a part of machine design, and it is my own practice to apply the principles of mechanics wherever I can do so, often when I know that it is unnecessary as a precaution against actual breakdown. This is done partly as a check of dimensions arrived at from other considerations, but mainly as a valuable *qualitative* analysis. For example, I regularly go through the analysis, before my classes, of the stresses called out in the frame of a Corliss engine by the effort transmitted from the piston to the crank. I cannot recall such a frame that seemed to be in danger of breaking from the actions which we can compute; but nothing is more suggestive as to the proper distribution of the metal in the frame—the form, not dimensions—of the sections than this analysis. In a punching machine the quantitative analysis is in itself valuable, and should always be applied.

Perhaps too much has been said of what we should teach, and the suggestions as to how to teach must be brief. The system I would suggest includes: Instruction in Kinematics (which I would preferably not divorce entirely from considerations of force), with abundant drill in drawing room work, and with other problems (graphical and analytical). A study of the physical properties of the materials of engineering, in the lecture room and laboratory. Lectures on the general functions and requirements of machines, with problems on the proportions of the elements and drawing room design. In conjunction with these, a carefully prepared set of exercises in sketching and measurement of actual machine parts, from the shops of the school and vicinity, should be given; and at times there should be exercises on "intuitive designing." For valuable hints upon carrying out this last named line of practice, I must again refer to the paper of Mr. Hawkins, already cited. Of the means available for bringing out the student's talents in handling those parts of design not amenable to algebraic treatment, one or two may be specifically given. Such parts include many of the principal dimensions, and nearly all of the details, or what may be called the filling, of the design. One method of developing the student's ability for this kind of work is to give the class the main dimensions of a simple machine member, after having discussed the general requirements of such a part in the lectures, and then to have the students sketch the complete member. For example, take a pillow block; give the diameter and length of bearing, the height above the floor to the center of the shaft, and perhaps the spread of the base at the floor. Or, for a lathe leg, give the width of lathe bed at the bottom, height above the floor, and spread of leg at the floor. I venture to predict that the pro-

duct of a class on such an exercise, will rival in variety
of form and finish the designs to be seen on the market.
This practice has two distinct and beneficial results;
one upon the student directly and the other upon the
instructor, and incidentally upon his instruction. The
student who finds himself confronted by this task with
no idea of what a lathe leg is like, will probably take a
good look at his lathe when he gets into the shop, and
if he does this, something has been accomplished. It
is too common for a boy to work in the shop for
months, without knowing his lathe by sight. The benefit to the teacher is in showing him wherein his instruction has failed to convey the idea he intended. One of
the hardest things for a man who lectures to a large
class is to determine what impression he makes upon
the individual student. An incident in my own experience illustrates this point. Upon one occasion the
class was told to sketch a pulley of given diameter, face
and bore. They had previously been told that a casting should be as nearly of uniform thickness as possible,
and that sudden change from a heavy to a light section
was to be avoided, in order to reduce the danger from
shrinkage. One of the pulleys showed the arms
thinned down at the rim to the thickness of the latter.
Opportunity was taken to clear up this misconception
at the next lecture; and it is hoped that this student
did not go out into practice and make a drawing of a
pulley which would tend to confirm some "practical"
superintendent in his belief that technical education is

a humbug, pure and simple. Beside the practice just discussed, other good exercises are such as the removal of some part from a machine or drawing, requiring the students to sketch a piece to take the place of the missing member. Another exercise is making a full size drawing of some part which is shown to small scale.

One other important matter must be touched upon; the equipment for instruction in design. Reference to the laboratories and shops has already been made, and these should be utilized in teaching design. Models— if good—are of service; but I believe that many of the models used are actually pernicious. Models are frequently employed to illustrate a kinematic motion, in which the members are made with utter disregard to good mechanical form. If these members are simply strips of wood, or wires (which are convenient in home made models), they may be so absurd as to be innocuous; but in many pieces of the more pretentious and expensive apparatus the form of the members is just bad enough to be insidious. The student failing to detect their faults, may carry away with him an impression which perverts his mechanical taste; the tendency to imitate a bad example is a human characteristic from which students are not entirely free. Actual commercial machines are the best models when these are not so large and heavy as to be seriously inconvenient. A small engine of good form, provided for adjustments, and properly sectioned, is an ideal valve motion model. To be sure, there are many motions shown in text-books of which it would be difficult to find examples in prac-

tical machines; and it is not at all certain that we can
afford to give much of the limited time of our course to
such mechanisms. The cost of practical machines for
models precludes their use in many cases, when it
becomes necessary to make the best practical substitute.
The very elegant set of Reuleaux models which we have
at Sibley College, contains many pieces of extreme
interest and great value in instruction, while other pieces
are seldom taken from the cases. Judicious selections
from this, and other similar sets, are desirable material
for teaching kinematics; I would say, if you have the
money for such purposes, choose your models carefully,
and see that what you get represents good practice in
all respects; but no teacher need despair of doing good
work because he has not the complete set.

One most valuable part of the equipment is a collection of trade catalogues, and any school can have
these. Among such publications of the present day are
the best treatises on machine design. In the work of
which the writer has charge (mainly steam engine
design), the students are encouraged to consult the catalogues freely, and to follow the practice shown by them
discriminately. In the design of an engine, each student is required to make the principal computations,
work out the steam distribution diagrams for various
loads, construct inertia and crank effort diagrams, etc.,
and then to make the general drawings and leading
details. He is advised to adopt some recognized type
of engine (or even make), or proper combination of
these, and only to depart from his type for good reason

in the design of special features. This is copying, in the sense in which architects copy classic construction. The forms employed in the practice of successful builders represent the survival of the fittest; and experience confirms the belief that a better training in design is secured by this system than by insisting upon strict originality in every detail. Occasional exceptions are of course to be made. A small quadruple expansion engine, for steam of 500 pounds pressure, has just been completed by two students in Sibley College. These young men have created that engine, designs, patterns and machine work. It is nothing if not original, and would be a credit to older and more experienced designers and builders. It is believed, nevertheless, that the practice outlined above is the better for general application.

Teachers in the larger and better equipped colleges unquestionably have certain advantages, while the brothers in the smaller institutions are not without their compensations. It the first place, the attractions of the well equipped laboratory make it difficult to create enthusiasm for the more prosaic work of design. Second, the numbers to be handled prevent giving much individual attention to the students in the larger schools. In the smaller colleges, owing to the absence of a complete laboratory equipment, less time is given, necessarily, to the experimental work, and more is available for drawing and design. One of the best trained designers I have ever had for a student came from a college whose graduates are not considered eligible for the advanced degree at Cornell; and he not only did superior work, but he needed little instruction. He came from an institution which has a small equipment, and only one man for all the work in mechanical engineering, but that one man is a man of genius, and a born teacher.

The instructor equipped with a good training in the schools, supplemented by a few years of general practice (this is essential) and with the true mechanical instinct, can do good work with little other capital. Let him collect his catalogues and as much else in the way of library, models, etc., as circumstances permit, and then throw his soul into his work. Success under these conditions means expenditure of energy and patience, but he may be encouraged with the thought that perhaps no branch of the mechanical engineering course is capable of yielding better returns on a small equipment, than is that of drawing and design.

THE EDUCATION OF CIVIL ENGINEERS FOR RAILROAD SERVICE.

By C. FRANK ALLEN,

Associate Professor of Railroad Engineering, Massachusetts Institute of Technology, Boston, Mass.

An inspection of the last catalogue of one of the largest technical schools of New England shows that one hundred and eighty-eight graduates in civil engineering may be considered to be now engaged in work where their education in civil engineering finds application. Of this number fifty-seven graduates, or thirty per cent. are engaged in some reasonable sense in railroad work. The capacities in which they are employed include those of president, manager, superintendent, chief engineer, subordinate engineer, bridge engineer, electrical engineer, road-master, supervisor of bridges and buildings, dealer in railroad supplies, and professor of railroad engineering. Bridge engineers, unless in the employ of some specified railroad company, are not included.

An inspection of the last list of "Members" of the American Society of Civil Engineers shows, from the addresses given, that two hundred and thirty-three are engaged in what may fairly be considered railroad work. The total number of members is one thousand, one hundred and seventy, but for very many of these, the

253

address gives no index of the character of their present
employment. The terms "civil engineer," "consulting engineer" occur often; and in many cases, the
city, street and number, constitute the only address.
Neglecting those whose occupation is not given, it is
found that the number of those whose address indicates
that they are not engaged in railroad work is four
hundred and forty-one. Of the total of six hundred
and seventy-four thus classified, the two hundred and
thirty-three engaged in railroad work constitute thirty-
five per cent. Many of those unclassified are known
to the writer as railroad engineers. Nevertheless, a
large proportion of them will, no doubt, belong to the
non-railroad class. As the two hundred and thirty-
three specified to be engaged in railroad work constitute twenty per cent. of the total membership, any
reasonable distribution of those unclassified could
hardly fail to show that those actually engaged in railroad work constitute as much as twenty-five per cent.
of the total.

Having in view the statistics of the American
Society of Civil Engineers and also the record of the
technical school referred to above, it seems probable
that of the graduates from a school who continue to
practice civil engineering, twenty-five to thirty per
cent. may be expected, at any given time, to be engaged
in railroad work. The facts being as we have found
them, the question arises whether there is sufficient
demand for railroad men with civil engineering education, to justify the schools in providing for those who

desire it, instruction intended to prepare them especially for railroad work, and differing from that given to other civil engineering students. In the opinion of the writer there is, and will be, sufficient demand. It is true, of course, that of those engaged in railroad work at any given date, some have previously been engaged in other branches of civil engineering, and some will not continue permanently in railroad work. To offset this, it is certainly true that there will be a greater demand for young graduates for railroad positions, whenever it shall be found that they have become better prepared to fill them. If it be urged that thorough discipline and drill in general civil engineering studies will furnish an ample foundation, so that the graduate will be able to adapt himself to any special work required of him, it may be answered that the same argument would largely hold in favor of a course of instruction, discipline and drill, in mathematics not immediately directed towards civil engineering. Just as it is found that the discipline and drill can be secured even more effectively for civil engineering students, when directed to civil engineering studies especially; in a similar way, the discipline and drill can be abundantly secured when especially directed towards railroad engineering, and with, of course, the advantage of direct preparation for the duties of the future. If, from any cause, it appears probable that specialization in this direction would result in a decrease of thoroughness, it should not be attempted. The discipline certainly is of primary importance. In engineering schools where the field is already too broad for the teaching force available, such specialization would be a mistake. The education of the engineer for railroad work should, in the opinion of the writer, be primarily an education in civil engineering, not in mechanical engineering. The important and difficult feature of the railroad is its construction, a critical part of which for successful operation, is the location, a matter peculiarly within the province of the civil engineer. In fact the location so far influences the difficulties and expense of construction, the cost of operating, and even the possible revenue, that it becomes strictly of primary importance.

If civil engineering be defined as it might be, in a modern limited sense, to be "the science and art of utilizing the forces of nature in producing fixed structures," then there would remain within the civil engineer's legitimate duties, among others, the construction of bridges, retaining walls and buildings, as well as track and the earthwork structure upon which it is in general supported; while the furnishing of equipment would lie outside his duties, being more distinctly a matter of mechanical engineering. The writer does not desire to underestimate the value of properly designed rolling stock for railroad use. It is, nevertheless, true that a locomotive of given weight, or a locomotive for given service, can be, and in practice, very commonly is, designed and constructed by one of the many companies formed especially for that purpose. In a fashion very similar, a bridge for a given span, and for specified loads, may be, and in practice, very commonly is, designed and constructed by one of the various bridge companies. It is not true in a corresponding sense, that a retaining wall can be designed and constructed, a pipe line located and laid, a station ground arranged, or the position of buildings determined, without, in general, a careful examination being made of the ground, often with the aid, it is true, of maps and profiles. In fact, so many and important are the civil engineering duties, as compared with the alternative, the mechanical engineering duties, that it seems clear that for railroad work, the proper education is essentially an education in civil engineering. The engineer for railroad work should be primarily a civil engineer, and secondarily a railroad engineer.

In what should this education in civil engineering and railroad engineering consist?

First, of preparatory studies, physics; free hand and mechanical drawing; and a thorough training in mathematics, including arithmetic, algebra, geometry, trigonometry, descriptive geometry, analytical geometry, calculus, and mechanics.

Second, of general professional studies, surveying, topographical drawing, stereotomy, strength of materials, bridges and structures, and hydraulics.

Third, of co-ordinate branches, physical geography, geology, metallurgy.

Fourth, of what may be considered general railroad subjects, some of these, as location, curves, grades,

17

earthwork, are essential; others, turnouts, spirals, a brief consideration of track, of train resistance, and of economics of location may reasonably be introduced as a part of general engineering training.

Fifth, of subjects adapted to students preparing specially for railroad work, economic theory of location, station grounds and yards, signals, train resistance, brakes and brake resistance, tunnels, track, street railways, building construction, arrangement of buildings; all of these are of a character suitable for instruction and discipline, and at the same time are matters regarding which the railroad engineer should be fully informed. In addition to these, railroad management, thermodynamics and the locomotive are subjects to which special reference will be made later. Comment or explanation is desirable regarding some of the subjects mentioned. What are here called general railroad subjects, location, curves, earthwork and others are not strictly railroad subjects although generally so classified. They properly apply to the location and construction of any engineering work where the length is considerable as compared to its breadth, so that the measurements and staking out are accomplished from a center line.

Of the subjects mentioned as special railroad subjects, the economic theory of location enforces strongly upon the student the necessity for the railroad engineer to build economically as well as safely, and to give due consideration to economy of maintenance as well as economy of construction, together with applications of this principle to railroad location. The introduction of systems of signals, and improvements in the arrangement of station grounds and yards, are destined in the immediate future to be of so great importance as to call for special instruction in those subjects. The consideration of tunnels furnishes a very favorable opportunity to call the attention of the student to the methods adopted in executing engineering work. The study of train resistance, or of brakes and brake resistance, is not only important in itself, but offers opportunity for theoretical training of the most excellent sort. Railroad engineers are almost sure to find it necessary to design and construct buildings, and should know enough of architectural construction to enable them to design, and order material for, buildings of the ordinary kind; they should also have their attention called to the arrangement of buildings so that they shall be convenient and economical for the special uses intended; round houses and coal chutes furnish instances. Street railways, too, have in late years attained an importance such that the subject should certainly receive attention.

The writer believes, as has been stated, that the engineer for railroad work should receive an education, primarily as a civil engineer, rather than as a mechanical engineer. In the operation of the railroad, questions as to the use of motive power are of constant, and of greater importance, than is the case in most of the other work where civil engineers are expected to exercise control. It follows that a thorough training and education as to the properties of steam and its application in the locomotive would prove to the railroad operating official of great value. Whether, when the requirements of civil engineering are satisfied in the course, there can then be found room for a satisfactory treatment of these subjects, and whether it will be practicable to furnish a limited course in the subjects fitted especially to the wants of railroad civil engineers, will in many cases prove a difficult problem, and in some schools would be rightly decided in one way, and in others in the other. The instruction would naturally be given outside the department of civil engineering. Whether the student would maintain an equal interest in what may be considered professional work outside of, and independent of, the work of his own department, especially in an abbreviated course, is again a question. The writer prefers at this time not to express positive views as to the advisability of introducing these studies, but this paper would hardly be complete if it failed to distinctly call attention to their importance.

The field for civil engineers in railroad work is not, and should not be, limited to the strictly engineering work. While the young engineer usually enters railroad service on the engineering side, he may be called upon to engage in work in other departments. In a lecture delivered several years ago, Charles Francis Adams divided the railroad into five great departments: 1. Financial. 2. Construction. 3. Operating. 4. Commercial. 5. Legal. We may adopt this classification as a convenient one.

The financial department is commonly under the immediate control of the president; and cases will be readily brought to mind in which the presidents of important systems are civil engineers. This is true of the Pennsylvania Railroad, for instance. Whether an engineering training furnishes the best education in preparation for the duties of this department may be doubtful. There can be no question, however, that the mathematical training received would be of especial value. Construction, of course, should be directly in the hands of the civil engineer.

The operating department, under the head of a general manager or general superintendent, requires for successful results, the services of men capable of acquiring a thorough grasp of practical and mechanical details connected with the roadway and equipment, and of the capabilities and defects of each. A technical engineering training furnishes the proper foundation from which practical experience and knowledge of detail can be most correctly and rapidly secured. The performance of the duties of operating railroads, has in the past, in many cases, been entrusted to civil engineers, and should be, and perhaps now is, considered legitimately within the lines of the practice of civil engineering, broadly considered.

The commercial department, dealing especially with railroad rates, properly requires at its head, a man of clear mathematical discernment, and preferably one who from his previous training will appreciate the effect of grade and other peculiarities of location upon the cost of operating, for this in the case of low class goods, certainly should form an element to be considered in fixing rates. A well devised course in civil engineering forms as good a preparatory education for such duties as can readily be secured. If we were to look for a master of the science of rate making, it would be difficult to find the superior of Albert Fink, a distinguished civil engineer, and the first Trunk Line Pool Commissioner.

In preparation for the legal department, the writer desires to strongly urge the claims of the technical school. The successful lawyer to-day is not the gifted orator, but rather that man who, with clearness of vision in analyzing his case, selects the point or points upon which success must depend, and who then upon these points carefully and thoroughly develops or constructs his case. The writer firmly believes that the analytical quality of mind is better developed by the mathematical studies, and the investigations required in our best technical schools, than in our colleges where the influence of classical education is still predominant. In the case of the railroad lawyer, especially, eloquence is of even less proportional value than ordinarily, in view of the fact that in well managed railroads no case is allowed to go to a jury trial, if any reasonable settlement can be otherwise secured.

If then an education in civil engineering is a suitable preparation for work in any of these five departments, should an attempt be made to teach such subjects as will qualify the student or graduate for duty in other departments, as well as he is now qualified for service as a civil engineer? The answer to this question must now be that a large proportion of students enter, and are for a considerable time engaged in, the work of civil engineering proper, and it would not be wise at present to take from engineering studies enough to constitute a really efficient preparation for the other departments. When this is granted, it does not follow that no attempt should be made to in any way prepare or stimulate the student to enter into these other departments of railroad service. The writer is not at present prepared to recommend more than a brief course in railroad administration or management, which shall acquaint the student with the important features of the railroad, its organization, methods of administration, and the economic questions involved in what is called the railroad problem, in such a way that the young engineer in railroad service shall be prepared to see and appreciate better what is going on around him, and thus broadened beforehand, to become the all around, efficient railroad man who will be in demand as important positions become vacant. Such a course would serve to stimulate and encourage graduates to accept positions if offered in the financial or commercial departments, or to supplement their engineering studies by a course in the law school. In time, it is probable that the demand for graduates of engineering schools would warrant the introduction of specially arranged courses to educate young men for general railroad service, with equal advantage to the railroad and to the engineering school.

The studies referred to as proper or desirable in a course in railroad engineering as distinct from civil engineering in general, would form a suitable course of study, or a major part of it, for graduate students who desire for any cause to supplement a general course in civil engineering already acquired.

As to methods of instruction to be pursued in educating civil engineers for railroad work, lecture, recitation, problem, fieldwork, drawing, design and memoir, can all be employed to advantage. Railroad curves furnish an application of mathematics acquired earlier in the course, and a thorough drill in the recitation room will usually be necessary to enforce properly and fix in the student's mind the mathematics which he thinks he has previously acquired. Most mathematical subjects are insufficiently mastered by students until attention is especially directed to them in connection with the applications. Problems should be given, and of such a sort that more than the mere substitution of values in formulas will be necessary. Again there is a tendency for weaker students to depend in greater or less degree upon stronger students in the same class; especially in the case of problems worked at home. The only method which will secure absolutely independent work is to have problems worked in the class and this should be done to as great an extent as circumstances allow. A systematic drill in laying out upon the ground circular curves, and transition curves or spirals, and in actually setting slope stakes, should form an important part of the fieldwork. For this purpose the class should be divided into parties of not more than four students. A railroad survey, two miles or more in length, should be made, in order to acquaint the student better with the details of fieldwork of this sort, to give point to his class work and yield additional experience in the use of instruments and in the applications of surveying. Whether the fieldwork should take place during the school year, or form a part of vacation work, is a matter upon which a difference of opinion may properly exist. There is some advantage in having the field-work and class-room work supplement each other. There is some disadvantage in having reconnoissance, preliminary survey, and location survey follow closely one upon the other, owing to difficulty in preparing the necessary drawings. There is also an advantage in alternating the severer studies with perhaps one day each week of fieldwork during good weather. In these particulars, fieldwork during the school year has advantages. On the contrary a better selection of ground will often be possible, there will be less of waste time going and coming, and there will be advantage in concentrating the attention and interest upon one subject, if the work be done during vacation.

Certain subjects must of necessity be taught by lecture. Where this is done, it should as far as possible be the object of the lecturer to enforce and illustrate principles rather than to simply recite facts or describe details. The student may occasionally be assigned a subject upon which he shall make preparation, and himself lecture to the class. So far as circumstances warrant, the student should by problem, design, recitation, or otherwise, be compelled to exert individual effort in a way not required in taking notes of lectures and later passing an examination. Problems and design dependent upon the lecture will often serve this purpose. The exercise in design need not always be made an exercise in drawing as well. Often a sketch design, not to scale, but with all necessary figures will serve the purpose as well or better, and with great saving of time. In some cases of course the drawing is necessary. A plan of proposed station grounds and yard, or a design for a system of signals will serve as illustrations of suitable problems in design. Subjects assigned students for investigation may be reported either orally in the form of lectures as mentioned above, or in writing, in the form of memoirs. A combination or alternation of methods, will, it is believed, more fully develop the student than a close adherence to any one method.

Success in railroad service does not result altogether from technical or engineering skill and acquirement. Energy and executive or business ability are elements of at least equal importance. In what way, if at all, does the engineering education prove of value in securing these requisites? To a considerable extent these qualities are acquired outside of and later than the formal education of the student. Yet it certainly must be true that the course of study laid down and required by our best engineering schools if faithfully carried out, does require on the part of the student an energy which is an earnest of future success. It is believed that the graduate of the engineering school, does, on the average, possess, in virtue of his course of study, an energy and earnestness lacking in the average college graduate.

Executive and business ability comes largely with experience in practical business life. Whether anything of great value can be accomplished in engineering schools in instruction in business methods of bookkeeping or accounting seems to the writer doubtful. Something certainly can be done to broaden the student and make of him at graduation something better than a mathematical engineering prodigy. Courses of study distinctly not professional, can be introduced to much advantage. Business law seems to the writer a very desirable subject for this purpose, and some knowledge of finance will certainly prove of advantage. Engineering societies organized and carried on by students will develop confidence in speaking in public, if there be no chance to give instruction in elocution. In many ways the student can receive instruction which will make it easier for him to rise equal to the emergency when the time comes. It should of course be borne in mind that a clear understanding of the business carried on is a very important element in the successful executive, and in this particular the engineering education performs its proper part.

Finally it can not be claimed that the engineering education is the only path to success in railroad service.

Promotion from the ranks will always be a means of reaching the highest positions. Nevertheless there ought to be a demand on the part of the railroads, for men with the education which our engineering schools can well provide, and our engineering schools on their part should stand ready to meet the demand, certainly as soon as this can be done without weakening work already undertaken.

ENGINEERING EDUCATION AND THE STATE UNI-VERSITY.

By Wm. S. ALDRICH,

Professor of Mechanical Engineering. University of West Virginia, Mor-gantown, W. Va.

The relation which should exist between these will be more clearly discussed if we understand the scope and meaning of engineering education to be that conveyed by the name and constitution of this society organized for its promotion. Engineering education is an education for a profession. As such its first requirement is a liberal education. This broad trend is best given by the pursuit of those studies affording mental discipline while developing a love of learning for its own sake and capable of giving an added grace to the exercise of future accomplishments. The professional education follows in course. Its distinctively technical features are to be combined with practical work of an educational value. Such will be found in selected exercises and activities from among those required in professional life,—in shop, office, field and laboratory. Engineering education while in the domain of science and an art is itself an art and a science.

ENGINEERING EDUCATION IN A UNIVERSITY.

With these and related considerations clearly before us it may be inquired whether any or all of them find a proper environment in a university. For the present purpose such an environment may be said to have a two-fold object: To produce and to sustain that variation of type which shall be most thoroughly adapted to continued existence in the given environment. If this is in a university, will there be any variation produced in the engineering education itself as well as in its product? If so, will it be desirable or undesirable? And whether one or the other, from what point of view must it be considered? Will engineering education thrive in a university atmosphere? Or, will it be frozen to death?

University presidents and renowned engineering professors have so placed themselves on record in this particular that we would fear the fate of any engineering training given within the walls of a university. But, what are the facts, as we know them to exist in the United States? Education for engineering as a profession has not only been recognized as entitled to, but has actually received and been correspondingly benefited by a university environment quite as much so as in the case of law and medicine. The liberal education required can now be obtained in a university as nowhere else. After finishing this part of his course, many believe that the student should continue in the university atmosphere while pursuing his technical studies and practical work. Many do so continue in state and other universities which offer engineering courses. With what result? Everything else being equal,—faculty, apparatus and all facilities for instruction,—these students are a different product from those who receive both academic and professional training entirely in a special school of engineering. This variation in the education itself as well as in its product, may be desirable or otherwise, according to the emphasis laid upon the academic compared with the technical training. Questions of finance as well as many others have been purposely avoided in considering the relation of engineering education to a university. It is necessary to arrive at some preliminary conclusions from first principles and known facts, and all irrespective of known disturbing elements. In no other kind of education will financial considerations enter to change so completely every condition and alter so entirely every product as in engineering education.

FEDERAL AND STATE AID TO HIGHER EDUCATION.

The extent to which such aid has been given is now a matter of history; and, educational history*

---

* "Contributions to American Educational History, No. 9."—The History of Federal and State Aid to Higher Education, by F. W. Blackmar, Ph. D., Bureau of Education, Washington, D. C., 1890.

that is so important and so vital to our subject that any presentation within the brief limits of this paper would not be satisfactory to all. The duties and responsibilities of the state in this direction have always been recognized, but appreciated chiefly in the line of provisions for academic training. The need for technical education began to dawn upon the people about the beginning of the century and received its first embodiment in the foundation of the Military Academy at West Point. Aside from this and the Naval Academy at Annapolis, however, technical education† received the most promising encouragement from private benefactions till its national endowment by the passage of the famous Land Grant Bill, July 2, 1862. It would be interesting, if time permitted, to examine into the conditions of national life and growth of educational ideas which led up to the passage of this bill‡— avowedly for the promotion of scientific and industrial education rather than what we have considered to be purely technical, much less professional from an engineering standpoint. It may be said that the profession of engineering, as we know it to-day, scarcely existed at that time. If it was so conceived as demanding any special education, still, legislation for the promotion of this would have been interpreted as legislation for a class at the the sacrifice of that for the masses.

---

† Technical Education in the United States, by Prof. R. H. Thurston; paper presented at the Chicago meeting (July, 1893), of the Am. Soc. Mech. Eng'rs, No. DXLIII., Vol. XIV. of the Transactions.

‡ History of the Agricultural College Land Grant Fund of July 2, 1862. Ithaca, N. Y., 1890. Publication of Cornell University.

THE LAND GRANT BILL OF 1862.

The "Colleges of Agriculture and the Mechanic Arts," established pursuant to the provisions of this bill, mark the beginning of a new period and of a peculiarly American development of national aid for promoting scientific and industrial education. In a few cases they were placed by the side of existing state institutions. Other states preferred to locate their new college near their geographical (if not political) centre, to allow its growth to be free from the traditional influences and environment of another and quite different system of education. This federal endowment was allotted in some states to new institutions or to form new departments of older institutions, and named after their principal benefactor. The original "grant" was a definite and permanent endowment to each state, of 30,000 acres of public lands, for each Senator and Representative in Congress. Such an apportionment was probably as equitable as could have been devised at the time; but, clearly very unequal in its vast possibilities. Aside from differences of population, at the time of the "grant," and the natural inequalities of statecraft, it still required only a few years to show the folly of some states and the wisdom of others in disposing of their "land scrip." The proceeds were for "the endowment, support and maintenance of at least one college where the leading object shall be, without excluding other scientific and classical studies, and including military tactics, to teach such branches of learning as are related to agriculture and the mechanic arts, in such manner as the legislatures of the states may respectively prescribe, in order to promote the liberal and practical education of the industrial classes in the several pursuits and professions of life."

There were the inevitable delays occasioned by the inception, organization and development of a new idea. Agricultural education languished; instruction in the mechanic arts had not yet developed into a science, and its incorporation into such colleges was more or less incomplete till endowed schools led the way; and, meanwhile, scientific, classical and military instruction carried the day. The degrees conferred were those of Bachelor of Arts and Bachelor of Science. However great the development has been along these lines,—and it cannot be denied but that it has been remarkable,—showing the wisdom and foresight of Senator Morrill in his connection with this "grant," still it will probably be admitted that this bill stimulated private endowments for technical schools quite as much as it directly benefited engineering education. Among the causes of such apparent failure, one has been noted; namely, the late period in the history of educational movements at which mechanic arts or shop training had become developed along lines at all educational. This period came about the time of the Centennial exhibition; but, it was too late for almost all of these land grant institutions. The important items of first cost and yearly maintenance confronted those that wished to conform to the requirements of the law in the matter of the federal endowment and

18

proceed to establish courses in the mechanic arts. The conservatism of the management and faculty of these institutions,—many of which were already living fully up to their income in maintaining the required scientific, classical and military instruction was a further obstacle to the introduction of such new courses.

THE MORRILL ACT OF 1890.

By this second national endowment, provided for by Act of Congress, August 30, 1890, engineering education, as such, is fully recognized, and some form of its development in all of the land grant institutions made possible. This endowment is from the proceeds of public lands which the federal and not the state government has the disposition of. Each state receives an equal amount in the form of an annual appropriation, irrespective of population or the possibilities of state development. From the first installment of $15,000, made June 30, 1890, it is increased by $1,000 each year, for ten years, after which it is to be maintained at $25,000. These radical differences in the amount and form of bestowal of the two national endowments are not without reason. The second is an effort at unification, at least in the apportionment of federal funds. The states feel the paternalism of the general government; but, that in which population and seniority play no part. The Morrill Act, of 1890, was for "the more complete endowment and support of the colleges for the benefit of agriculture and the mechanic arts" established under the provisions of the Land Grant Bill, of 1862. There are only two ways to realize this: (1) by increasing the salary account; (2) by providing additional facilities for instruction, such as apparatus, machinery, text and reference books, stock and material.

It has been found that the salaries of certain chairs, formerly paid out of state appropriations, could be paid out of the Morrill Fund, because the instruction given by the occupants of these chairs was within that provided for by this Act. Finding that certain salaries could be drawn from this fund did not carry with it the obligation that they should be so paid. It has developed state parasitism in a very unexpected manner. It frustrates the very intent and meaning of the Morrill Act, for the land grant institutions of such states are not thereby more completely endowed. The actual endowment remains practically the same; it is increased from Washington but decreased from the state capital to the extent of those salaries so paid. This Act was clearly not made a law for the purpose of shifting salaries or any other expenses, off the state budget and onto the Morrill fund,—however legitimate such disbursements might be. It is the state treasury or another institution which thereby receives "more complete endowment and support." In the same year that this was done hundreds of thousands of dollars were appropriated by the legislature of one and the same state for an Insane Asylum! Pounds of cure; ounces of prevention. Will not our states suffer from this untoward and unanticipated development of parasitism? In particular, will not engineering education lack the encouragement and development it was destined to receive from this Morrill Act ?

"The more complete endowment and support" of engineering education is most emphatically provided for. Aside from instruction in agriculture and the English language, the remaining branches specified in this Act are: "The mechanic arts, * * * * and the various branches of mathematical, physical, natural and economic science, with special reference to their applications in the industries of life, and to the facilities for such instruction." It seemed destined to endow and maintain courses in the mechanic arts, at least. Engineering laboratories appeared also in view; and some portion of the annual appropriation might be allotted for maintenance of instruction in this newly required branch of experimental engineering. If the former Act, of 1860, was ahead of its time, in seeking to provide for mechanic arts instruction, surely the latter Act, of 1890, was not at all so anticipative of the next development of technical education,—that of experimental engineering. This was waiting for an endowment with which to get a start in the original land grant institution.

Engineering education is particularly crippled in those states where parasitism has developed in the Morrill fund. In others agricultural education receives the lion's share of the apportionment of this fund, which it will be shown has been indirectly provided for by another federal endowment, at least so far as material equipment of work rooms and laboratories is concerned. It is just in this direction that engineering education would receive the needed encouragement and support from the Morrill Act; and, not only to develop new lines of such instruction but to further endow the established courses of land grant colleges and universities that they might have an air of respectability among institutions of their class in other states. After providing liberal education for the young engineer the next duty of the state institution is to apportion its resources and adapt its facilities for his technical training so as best to prepare him for the particular professional demands that will inevitably be made upon him in his own state. Now the parasitism, in reference to the Morrill fund, is not only most likely to occur but actually has occurred in just those state institutions in which some such development of old and beginning of new engineering courses had long been deemed the necessary first step towards educating their own sons to develop the state's resources and share in the promotion of its industrial progress.

Another feature of this Morrill act will affect the promotion of engineering education in every state to which the particular provision applies. Wherever a state makes any distinction of race or color in the admission of students to its land grant institution, there must be "a just and equitable division of the fund to be received under this act between one college for white students and one institution for colored students." The policy of this provision it is not the purpose of this paper to discuss. It will be sufficient to state it and draw some inevitable conclusions. It means two faculties, a double equip-

ment and a separate establishment throughout for the
white and for the colored students.   It is in just those
states which need, for developing engineering education,
all that the Morrill fund will legitimately provide, that
such a division of it will be felt most keenly.   If they
wish to establish and maintain reputable technical
courses; to provide recognized facilities for practical
work and experimental engineering to justify their sons
in seeking professional training in their own state institution rather than elsewhere, by offering an engineering education that will be at all comparable to that of
other state institutions;—the legislatures of these
states will need to make additional appropriations.   In
some of them it will be all the more difficult by reason
of the parasitism that has developed already from a
spirit of retrenchment in educational appropriations.

Federal aid for the promotion of engineering education is fraught with dangers as well as possibilities.
But, aside from the above considerations, it is much
less promoted by the Land Grant Bill, of 1862, than by
the Morrill Act, of 1890, for two principal reasons: (1)
The classical and military instruction, provided for by
the former, are excluded by the latter; (2) Instead of
the unequal permanent endowments of the former there
is an equal annual appropriation given to each state by
the latter.   The effect of this favorable difference has
been felt already in several of the states.   It has stimulated private benefactions for the promotion of engineering education.   It came, also, at a time when
urgently needed to assist in establishing new and rap-

idly growing engineering courses and aid in the material
equipments for the same, such as those in electrical
engineering.

STATE AID TO ENGINEERING EDUCATION.

The least the state could do, in accepting the federal endowments, was to meet the simple requirements
of the law,—to purchase, erect and maintain suitable
buildings in which the instruction provided for by the
general government could be materialized. The initial
and other building appropriations, as well as such made
for increasing the facilities for instruction and providing for continued effective maintenance, have placed
some of the original land grant colleges and universities in the very front rank among our state institutions.

STATE UNIVERSITY LIGHT, HEAT AND POWER PLANTS.

The establishment of light, heat and power plants
by some of the states in their institutions has rendered very material aid to engineering education.
These localize the generation of light, heat and power
in a central station from which distribution is made to
all parts and departments of the institution. These
were formerly urged from business reasons alone, such
as securing greater economy of installation, maintenance and supervision. It was soon made evident
that it would be a great advantage to have all the equipment of such a plant gathered about the engineering
shops and laboratories, increasing still further the
economy of repairs, maintenance and supervision. Not
only so, but the whole equipment thereby becomes
available at any time during the college year for experimental engineering work; and all of this could be

conducted on a more practical and commercial scale
than if simple experimental machinery had been alone
installed for such instruction. The state would thereby
help itself reduce the current and contingent expenses
for light, heat and power; and at the same time provide
admirable facilities for shop and laboratory training in
electrical, steam and hydraulic engineering entirely
beyond the range of the ordinary apportionment of the
Morrill fund for such practical instruction. In this
direction state pride and competition will enhance still
further the great economic and educational value of
such plants. American university teaching is coming
to require light, heat and power as necessary facilities
for the most efficient prosecution of laboratory methods
of instruction. The new university may come to bear
to the old, in this matter of material equipment, the
relation that a modern man-of-war bears to an old
time line-of-battle ship.

AGRICULTURE AND MECHANIC ARTS.

These have been inseparably connected in the
minds of statesmen when planning and developing federal aid for promoting such instruction. Should the
national endowments continue to be given jointly or
severally for these branches? It will be seen that this
has depended upon educational considerations, on the
one hand, and research, on the other. By the acts of congress of 1862 and 1890, federal aid is given for the joint
"endowment and support of the colleges for the benefit
of agriculture and the mechanic arts." By the act of
1887, each state sustaining such a land grant institution

.receives $15,000 annually for the establishment and maintenance of an agricultural "experiment station." When it comes to the question of federal aid for the promotion of scientific research, however, statesmen are divided. There is no engineering "experiment station" yet endowed by the general government.

Why should agricultural research be considered so proper a subject for federal aid and engineering research scarcely dare to beg for such recognition? It is said that agricultural education, as such, pure and simple, has too often languished, and, in many institutions almost died out; that the only substantial way to benefit scientific agriculture is by the federal endowment of "experiment stations;" that agricultural education will be the most promoted thereby, even if indirectly; and, that it will surely rally about these, if at all. The general results have been much as were anticipated.

The rapid growth of experimental engineering within the last decade; its recognized value for the determination of engineering data and precedents; its incorporation into the professional courses of almost all American technical schools,—bespeak a like consideration for engineering "experiment stations," or laboratories, in connection with all land grant institutions. For, notwithstanding that engineering education owes a very large part of its present high state of development to private munificence, still, the work now done in its shops and the researches carried on in its laboratories would be greatly accelerated by such federal aid as has been given to agriculture.

STATE ENGINEERING "EXPERIMENT STATIONS" OR LAB-
ORATORIES.

The technical equipment of such laboratories would
be of invaluable service in engineering instruction; and,
this in addition to that of the light, heat, and power
plant, which it is the duty of the state to establish.
Engineering instruments would be standardized and
tests of power plants conducted by disinterested parties.
Engineering practice throughout the state would be
reciprocally benefited; state laws relating to boiler ex-
plosions and engineer's licenses subject to careful
supervision; and, the duty of the state in protecting
life and property from engineering accidents, casualties
and negligence receive attention commensurate with that
now given by the agricultural "experiment stations,"
to healthful foods, economic forestry, farm and dairy
sanitation, and stock raising. The state's resources of
materials for building and other constructive work, as
well as fuels,—the prevention of their waste and the
utilization of by-products,—would receive careful consideration. Such a state engineering laboratory should
be allowed the same immunities and granted the same
privileges and opportunities as the agricultural "experiment station" now enjoys, in properly charging for
tests, researches, analyses and other scientific investigations. But, beside this, it should receive state aid for
the publication and interchange of bulletins, quarterly
or oftener; and, this would add as much to the value
and permanency of its year's work in engineering as is
now done for agricultural interests. The accumulation,

classification and preservation of engineering literature; records of tests, researches, and other investigations; and data for engineering practice and precedent;—all would receive an amount of attention that it is almost impossible for individuals, corporations or manufacturing establishments to give. What has been done by private munificence, in many of the above lines, in a few of our technical schools, should be repeated, extended and made possible, by federal and state aid, in every state college and university.

STATE AND ENDOWED INSTITUTIONS.

There are three classes of institutions in which engineering education receives more or less attention:

(1) Those dependent entirely upon federal and state aid.

(2) Those receiving private endowments in addition to federal and state aid.

(3) Those dependent entirely upon private endowments.

For any particular branch of engineering education in any one of these institutions the technical studies of the class-room will be found much the same, while the academic requirements differ widely. The kind and amount of practical work in office, field, shop and laboratory will vary according to the nature and extent of the endowment.

State institutitions are created and maintained by the people. They exist for the greatest good to the greatest number of students. Candidates who are unable to enter the freshman class cannot be turned away. In some such institutions high school work must be done, in providing preparatory courses for those applicants who have not had the advantages of city life. With few exceptions, tuition is free in all departments to all state students. In general, the age of the students is greater than in endowed institutions. All those who pass the entrance examination must be received, whether faculty and facilities are at all commensurate or not; if both are alike insufficient to provide for the great number of students admitted,—there is no better argument with which to go before the state legislature for additional appropriations. However technical the instruction may be, it is necessary to bear always in mind the particular needs of the state in whose institution it is given. Students of state institutions often enter for one year only, to try the work, and get as much out of it as their resources will permit. Others enter for two years. Very few enter for a definite four years' course clearly in view and fully prepared for financially. While this entails a growing responsibility on engineering teachers in all state institutions it is likewise an opportunity. More elementary technology in the first two years would open the young man's eyes, arousing an ambition to remain, or perchance to return after an absence of a year or more. It would help the student better to earn a living during such absence occasioned by financial stress. More practical work before leaving would also relieve him of its equivalent after returning. Thus to leave his educational work for a year or two is not the worst thing that can happen to the young engineering student. In the state institution, also, methods of study, instruction and discipline are likely to be quite different from such in endowed institutions. Many of the engineering students will be state cadets, receiving the military instruction and training provided for by the Land Grant bill, of 1862. This makes possible an organization and development of practical work in engineering along such lines as have been worked out at the Naval and Military Academies, in detailing students for the conduct of engineering work, in office, field, shop or laboratory.

Endowed institutions are mainly supported by incomes from first endowments and by tuition fees. They exist for the good that they are able according to their several abilities to give to those who can pay for it. Therefore a high standard of admission may be set and maintained. If in good financial and professional standing they will take only that number of students best adapted to the faculty and facilities for instruction. All students enter for the regular four years' course, practically by competitive examination. The technical instruction may be carried to any ideal standard, for no respects are required to be paid to any one community or class of interests more than another, — insuring a purely professional course, with the corresponding degree. The influences, shaping the character of engineering education in our state institutions have produced a certain uniformity when operating from without; but, a variety of type, if from within. Those from without, such as the federal aid of 1862, resulted in a somewhat uniform development of science teaching, while that of 1890, may do the same for engineering instruction. Shops and laboratories are being built with great rapidity and with more or less uniformity in equipment and plan of organization. The newly added engineering courses are usually organized and directed by graduates of the few leading endowed technical schools and of the Naval and Military Academies,—developing the new along lines of the old. On the other hand, such influences from within as the geographical location of the state, its natural resources, the traditions and customs of the people, their present and prospective industrial life,—all tend to shape engineering education so that it may best fulfill its mission in its own state rather than in any other.

STATE COLLEGES AND UNIVERSITIES.

In the state colleges of Agriculture and the Mechanic Arts, it might appear that engineering education would receive a higher development than in state universities. In the former the funds are not apportioned among so many different departments as in the latter, whose aim it only too often is to so multiply departments and courses as to be worthy at least of the name of a university. Each institution has its peculiar advantages and disadvantages from the point of view of engineering education. The state college is distinctively a school of science. Its courses are arranged, its equipments selected and its faculty appointed with this one object in view. In this respect it approaches the position of the endowed school of science. To add an

engineering course, in such, is to introduce its purely
technical features. The state university has a greater
variety of interests to serve. Here, science teaching
has to meet, of necessity, the varied requirements for
degrees in many different and often unrelated courses.
If the resources of the university are small, one professor and one course are made to cover the ground for
all the science work. In such institutions the science
teaching required as a basis for engineering is likely to
be unsatisfactory and the technical courses correspondingly weakened. In both institutions the old
academic courses are abridged to make room for the
new engineering studies. The effort is made so to
combine the essential requirements of a liberal education with sufficient technical training as to warrant
conferring the degree of Bachelor of Science in Engineering,—a combination peculiarly significant of the
development of engineering education in land grant
institutions. As both state college and university
receive equally from the Morrill fund, it would appear,
other things being equal, that the college student may
in time have a better faculty and larger equipment with
more specializing facilities to work with. Some will
argue, however, that such material advantages will be
more than offset by the student lacking completely the
liberalizing tendencies of the university environment.

Are these conditions such as should exist in state
institutions? Should some of them endeavor to cope
with other institutions which receive private endowments beside their federal aid? Or, even to equal

privately endowed schools of engineering? Is it advisable to announce the same course, seek to provide a correspondingly competent faculty, endeavor to maintain the same equipment, and give, at the end of four years' work, the same professional degree? An investigation of our state institutions will show that almost all of them give the degree of Bachelor of Science in engineering, at the end of four years and the full professional degree on the completion of one more year of resident study. A few, however, give the full professional degree at the end of four years. The courses, faculty and equipment of these should be carefully compared with those of endowed schools to arrive at any just appreciation of the intrinsic value of the professional degree conferred.

FACULTY ORGANIZATION FOR ENGINEERING EDUCATION.

The introduction of engineering education into a state college or university,—whether prompted by the ambition of the institution or demanded by the people, will require an organization pursuant to the Morrill Act which will be effected by an infusion of new ideas, new methods, new men and new appliances. It cannot be otherwise and succeed. For the governing board to direct the president to distribute the studies among the several professors is to develop such an institution on the plan of a country school, where any man may be expected to teach anything at any time. In the evolution of engineering education in the state university, three distinct stages of organization of the teaching staff may be recognized:

(1) Engineering teachers are members of one common faculty under one president.

(2) They are organized as an engineering faculty similar to the law and medical faculties which usually exist at this period of development.

(3) The organization of the engineering college within the university, with its dean or director, whose duties are related to this college as those of president to the university.

The faculties of state institutions are much less stable than in endowed schools. Aside from political causes the reasons are obvious and well known: uncertainty of tenure of office due to successive appointments for one year only; having to tutor in one department while head of another; having to assume half the work of a vacant chair for one year, as a temporary expedient for economy's sake, which is not relieved but becomes a precedent for succeeding years; distributing work among engineering teachers after the manner of the academic staff, any one of whom might be required to teach Latin, Greek, or modern languages, mathematics, history, literature or philosophy, as the occasion of each year demanded or the economics of administration dictated. There are engineering ethics and equities in teaching as well as in the practice of the profession.

Marked changes are taking place in the organization of all educational work. Shall we not have subdivision of labor and consequently of responsibility in

19

this field when it has become so fully recognized and established in all other lines of human activity? The most efficient professional teaching can be done and the most highly developed product obtained by specialization of supervision and thorough departmental organization. The organization of the engineering faculty is the first step in the right direction. The president of the university presides at all of its meetings; in his absence, the senior professor of engineering. The need for some such organization in state institutions will be apparent from a careful examination of the existing forms of faculty organization in many of them. The engineering college is a development falling into line with a plan of university organization which has been remarkably successful in other branches in European universities. The organization of schools and colleges within the university has been fully tried and found well adapted to the management of a great institution. Engineering education in our state universities admits of such an ultimate ideal organization. Few of them have already taken the lead.

ENGINEERING EDUCATION IN A NATIONAL UNIVERSITY.

The national endowments, of 1862 and of 1890, may alone be sufficient to enable several state institutions to get a start in at least one branch of engineering education. It may be regretted that this is the only source of income for such education in many of them; still, it has been shown that very substantial aid may be rendered by the state in furnishing buildings and in establishing a light, heat and power plant for its institution. On the other hand, if these federal endowments are diverted into other channels: serving to maintain several courses in a university instead of a few in a college; or, for an unusual development of agricultural education; or, to pay salaries formerly paid out of state appropriations; or, to sustain an institution of like grade for colored students;—then the state appropriations for engineering education will require to be proportionately increased. A new national endowment for engineering "experiment stations" or laboratories, to be established in all land grant institutions, after the manner of the agricultural "experiment stations" (provided for by the act of 1887), would supply the greatest need of all such institutions;—namely, facilities for experimental engineering; and, this in addition to the other reason which has called such "experiment stations" into existence: the value of federal aid for the promotion of independent scientific research.

The federation of all state colleges and universities into a national university, with its educational centre at Washington, would place such an institution in the very midst of the most favorable environment for the prosecution of that kind of advanced educational work for which it came into existence. A select committee unanimously approved and recommended for passage (March 3, 1893), the senate bill 3824, reporting as follows:

"Such an institution only could in any proper sense complete the now incomplete system of American education and most wisely direct all worthy efforts in the field of original research and utilize the facilities for it so rapidly accumulating at Washington. It provides for the establishment of a university of the highest type, resting upon the state universities and other institutions of collegiate rank as they rest upon the high schools and academies;—a university whose facilities shall be open to all who are competent to use them; but, whose degrees shall be conferred upon such only as have already received a degree from some institution recognized by the university authorities; * * * and, whose several heads of departments are to have advisory and co-operative relations with the heads of government bureaus for the mutual advantage of the government itself and the cause of universal science."

Library
N. C. State College

www.ingramcontent.com/pod-product-compliance
Lightning Source LLC
Chambersburg PA
CBHW021507210326
41599CB00012B/1169